THE BUSINESS OF BIOTECHNOLOGY

Profit from the expanding influence of biotechnology

Third Edition

Yali Friedman, Ph.D.

LOGOS PRESS

THE BUSINESS OF BIOTECHNOLOGY

Third Edition

by Yali Friedman, Ph.D.

Published in The United States of America
by
Logos Press, Washington, DC
WWW.LOGOS-PRESS.COM
INFO@LOGOS-PRESS.COM

Logos Press is an imprint of thinkBiotech LLC
Copyright © 2008, Yali Friedman, Ph.D.

Third edition

10 9 8 7 6 5 4 3 2 1

ISBN-13:   978-1-934899-00-7

Contents

Introduction

Science

Laws and Regulations

The Business of Biotechnology

Conclusion

Appendices

Figures and Tables

Preface

It is absolutely essential to recognize that success comes
at the end of failure after failure after failure … If it were
easy, 500 people would have already done it.
Alejandro Zaffaroni

The biotechnology industry was born in 1973 with the development of gene splicing techniques, enabling the directed modification and use of biological systems. The potential to improve drug development methods was quickly recognized as many companies were formed to leverage gene splicing and other related techniques and discoveries. Biotechnology has already altered paradigms for drug development, improving efficiencies and enabling new possibilities, and its influence is expanding to revolutionize other industries.

New discoveries and technologies continue to expand and extend the applications and appeal of biotechnology, enabling those in previously unrelated fields to profit from the expanding influence of biotechnology. Biotechnology has improved crop yields, drastically reduced the need for pesticides, and created nutritionally-enhanced foods. Biotechnology also shows great promise for industrial processes: increasing efficiencies of existing processes, reducing waste generation, and enabling unprecedented new possibilities.

This book presents a history and overview of the biotechnology industry, explaining the scientific, legal, regulatory, and commercial factors that shape and define the industry. Armed with this understanding of the drivers of the biotechnology industry and an appreciation of the applications of biotechnology, you are equipped to profit from the expanding influence of biotechnology.

This book is presented in five sections: a general introduction; the science of biotechnology; legal and regulatory issues; the business of biotechnology; and a conclusion. The scientific, legal and regulatory issues are presented prior to the business fundamentals because in order to understand the business of biotechnology it is necessary to first understand how the convergence of these factors defines the biotechnology industry. The final section includes investment and career development strategies. A comprehensive set of appendices follows, containing Internet links, an annotated bibliography, and a detailed glossary.

Several special considerations have been included to promote accessibility. Individual biotechnology companies and products are referenced in different examples and anecdotes to reinforce the concepts presented. Extensive cross-references are also included throughout the text for those readers taking a 'cafeteria-approach' and reading the chapters out of sequence. The annotated bibliography and detailed glossary facilitate continued learning by interested readers.

I hope that by breaking down the biotechnology industry to its key drivers and by providing numerous case studies, you will develop an appreciation of the independent and combined scientific, legal, regulatory, political, and commercial influences that define the scope of commercial biotechnology.

                                                    – Yali Friedman, Ph.D.

I

Introduction

Chapter 1
Introduction

> The ability to manipulate the genetic codes of living
> things will set off an unprecedented industrial
> convergence: farmers, doctors, drug-makers, chemical
> processors, computer and communications companies,
> energy companies, and many other commercial
> enterprises will be drawn into ... what promises to be
> the largest industry in the world.
> *Juan Rodriguez and Ray A. Goldberg, March 2000*
> *Harvard Business Review*

Biotechnology inventions and products are changing
paradigms in healthcare, agriculture, and industrial
processes. Great opportunities exist for those who have
the technologies, skills, and perseverance to bring new biotechnology products to market. These opportunities stem from the
disruptive effects of biotechnology on existing markets (and
the ability to create new markets), but they are tempered by a
unique set of scientific, regulatory, political, economic, social,
and commercial influences. Understanding the dynamic and
linked contributions of the interdisciplinary array of factors affecting the commercialization of biotechnology is essential to
operate in the biotechnology industry.

The biotechnology industry is not defined by a set of products or services, but by a set of enabling technologies. Whereas
the literal definition of biotechnology encompasses everything
from traditional agriculture to soap-making, modern defini-
tions describe applications relying on more complex and sophisticated techniques such as genetic engineering and other

3

Table 1.1 What is biotechnology?

Product / Service	Description
Biodegradable plastics	Reduce environmental impact of consumer goods
Diagnostic tests	Determine human predispositions to disease
DNA analysis	Determine paternity and assist in forensics
Genetic testing	Assist in traditional plant and animal breeding
Genetically modified crops	Improve yields and nutritional properties
Industrial enzymes	Improve efficiency and reduce environmental impact of industrial processes
Therapeutics	Treat and cure diseases

forms of directed modification of living things. This book defines biotechnology as the application of molecular biology for useful purposes. This distinction is important, because whereas inclusion of traditional activities describes processes with established markets and mature technologies, the focus on modern techniques reflects the innovative and revolutionary possibilities of molecular biology: manipulating living organisms and parts of living organisms to capitalize on scientific discoveries, to improve upon existing solutions, or to serve new markets.

Biotechnology has applications in health, agriculture and farming, environmental remediation, and industrial processes. Within the diversity of biotechnology applications, there are two basic modes of development: products and services. Certain drugs, such as those produced in bacteria, yeast, and mammalian cells, are examples of biotechnology products (the distinction between biotechnology-derived and traditional pharmaceutical drugs is discussed in greater detail in Chapter 4). Drugs, and biotechnology research tools that are sold to pharmaceutical and other biotechnology firms, are also examples of products. Services can be sold to research firms or to companies further down value-chains for downstream application. Genetic testing is an example of a biotechnology service and is used to determine parentage, to resolve identity issues in criminal cases, and

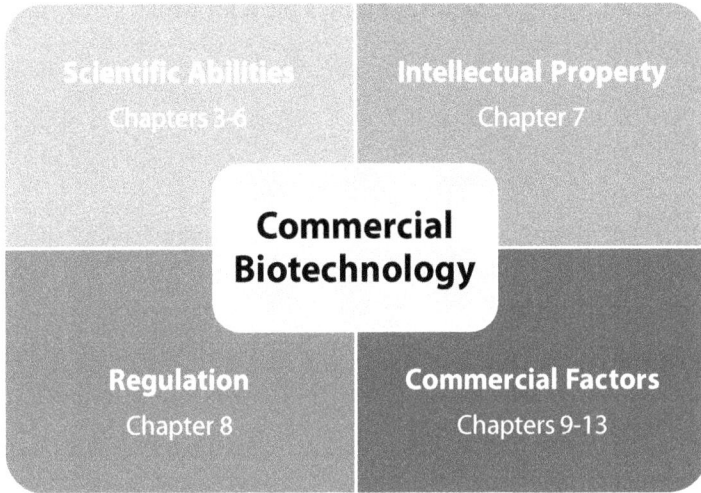

Scientific Abilities Chapters 3-6	Intellectual Property Chapter 7
Commercial Biotechnology	
Regulation Chapter 8	Commercial Factors Chapters 9-13

Figure 1.1 *The four pillars of biotechnology*

to screen for predispositions to disease.

The possible applications of biotechnology are defined by current scientific knowledge and abilities, and by the capacity of companies to develop marketable solutions from current knowledge or through additional research. The commercialization of biotechnology applications is further promoted and limited by numerous legal and regulatory factors. Patents serve both as a barrier to entry and an incentive for development. Changes in patent law can have profound implications on the ability of biotechnology firms to operate profitably and to obtain financing. Approval from bodies such as the Food and Drug Administration, the Department of Agriculture, and the Environmental Protection Agency is also required before many biotechnology products can be marketed or even tested.

Beyond these fundamental factors which define the possible applications of biotechnology, commercial factors also play an important role, as biotechnology ventures must ultimately be profitable. Whether structured as a for-profit company or a non-profit entity supported by donations or government grants, any biotechnology venture lacking an income stream cannot be sustained. Survival requires filling a need for which some party is willing to pay.

Chapter 2

The Development of Biotechnology

In science the credit goes to the man who convinces the world, not the man to whom the idea occurs first.
Sir Francis Darwin

The modern biotechnology industry is built upon knowledge and techniques developed in the pharmaceutical industry, which employed biological extracts, dyes, and complex organic and chemical mixtures to produce drugs.

The emergence of the pharmaceutical industry is partially attributed to the development of aspirin, a drug that was developed by the German industrial chemist Felix Hoffman in 1897 and is still commonly used today. Many patients, including Hoffman's father, could not tolerate the stomach irritation associated with sodium salicylate, the standard anti-arthritis drug of the time. Armed with the knowledge that acidity associated with salicylates caused stomach discomfort, Hoffman sought a less-acidic formula and eventually produced acetylsalicylic acid, or aspirin.

As medical knowledge advanced, a focus on symptom-based treatment of diseases replaced techniques such as bloodletting and led to research on the effects of medicines and the use of defined substances as drugs. The emergence of a rational basis for medicine supported research on human biology based on the belief that a better understanding of human biology would lead to better medicine. At the same time, improved knowledge of microorganisms related to human health led to an understanding of the causes of infectious diseases and allowed new treatment paradigms. Penicillin, for example, was identified as a potential anti-infective drug based on the observation of its ability to prevent the growth of bacteria in laboratory experiments.

The growth of the pharmaceutical industry paralleled advances in knowledge of general biology and advances in methods to study and manipulate biological systems. The emergence of refined tools permitted a more fundamental study of biology—molecular biology—focusing on the fundamental processes affecting biology. The discovery of the structure of DNA in 1953 was instrumental in developing an understanding of how genetically inherited characteristics are passed from generation to generation.

The first biotechnology companies were formed in the 1970s and 1980s. Knowledge of the molecular fundamentals of biology and development of tools to manipulate biological systems laid the foundation for the biotechnology industry, which employs the directed application of molecular biology for useful purposes. Biotechnology drug development not only uses methods and strategies different from traditional pharmaceutical development, it also produces different products. By selecting proteins such as insulin and erythropoietin, whose functions were already known, as their lead compounds, firms such as Amgen, Genentech, Chiron, and Genzyme employed a directed drug design strategy. In contrast with the chemical synthesis and biological extraction techniques that produced traditional pharmaceutical drugs, these early biotechnology companies used recombinant DNA techniques that enabled them to produce proteins as therapies (see Chapter 4 for more details).

KNOWLEDGE AND SKILLS

A brief history of selected Nobel Prize awards in the categories of Chemistry, and Physiology or Medicine provides a path to follow the scientific developments that spawned the biotechnology industry. Nobel Prizes are awarded for outstanding achievements and contributions and are internationally recog-

1940s	1950s	1960s	1970s	1980s
· Protein sequencing	· DNA structure · X-ray crystallography	· Genetic code	· Restriction enzymes · DNA splicing · DNA sequencing · First biotech companies	· Polymerase chain reaction · First biotech-derived drugs approved

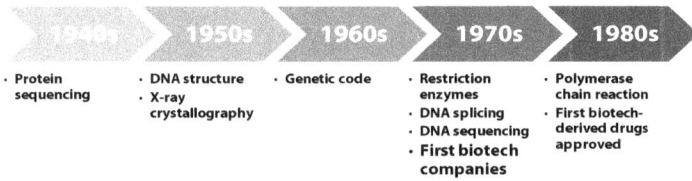

Figure 2.1 *Knowledge and skills enabling biotechnology*

nized as the most prestigious awards in the fields for which they are awarded. Because it can take some time for the significance of a discovery to emerge, many Nobel Prizes are awarded years after the actual discovery.

Frederick Sanger was awarded the Nobel Prize in Chemistry in 1958 for his determination of the protein sequence of insulin. Sanger, who began his mission in 1943, developed numerous techniques to directly sequence proteins, which enabled scientists to better understand these biological molecules. Knowledge of the sequence of human insulin enabled Genentech to develop recombinant human insulin—the first biotechnology drug—in 1982.

Between 1950 and 1956, Herbert Hauptman and Jerome Karle laid the foundations for the development of X-ray methods to determine the structure of crystallized molecules. They shared the 1985 Nobel Prize in Chemistry for their work. X-ray crystallography determines a molecule's three-dimensional structure by analyzing the X-ray diffraction patterns of crystals of the molecule. The complexity of organic molecules such as DNA and proteins meant that many structures were not known until the advent of X-ray crystallography. X-ray crystallography aided discovery of the structure of DNA in 1953, a significant advance in molecular biology that set the stage for modern biotechnology. James Watson, Francis Crick, and Maurice Wilkins shared the 1962 Nobel Prize in Physiology or Medicine for their work in discovering the structure of DNA. This discovery enabled elucidation of the mechanisms for control of gene expression and hereditary transfer of genetic information.

Following the discovery of the structure of DNA, the need to explain its role in cellular functions remained. Robert Holley, Har Gobind Khorana, and Marshall Nirenberg shared the 1968 Nobel Prize in Physiology or Medicine for their contributions in deciphering the genetic code, the language by which information is contained in DNA, and for elucidating how this information is translated by cells.

Werner Arber, Dan Nathans, and Hamilton Smith shared the Nobel Prize in Physiology or Medicine in 1978 for the discovery of restriction enzymes and their application to problems of molecular genetics. It was the pioneering work of these three scientists that enabled development of the DNA manipulation techniques that permitted Stanley Cohen and Herbert Boyer to develop methods for splicing DNA from different sources, often referred to as recombinant DNA (rDNA) technology.

The 1980 Nobel Prize in Chemistry was awarded to Paul Berg, Walter Gilbert, and Frederick Sanger. Berg was recognized for his "fundamental studies of the biochemistry of nucleic acids, with particular regard to recombinant-DNA," and Gilbert and Sanger for their "contributions concerning the determination of base sequences in nucleic acids." The ability to determine the sequence of DNA was central to the Human Genome Project and is a key element in biotechnology research and development.

Kary Mullis and Michael Smith shared the 1993 Nobel Prize in Chemistry for their respective development of the polymerase chain reaction (PCR), and site-directed mutagenesis. Mullis' PCR permits the specific production of copies of a specific DNA segment, even in the presence of a complex mixture of DNA. This technique has applications in forensics, paternity and heritage testing, medical diagnostics, archaeology and anthropology. Application of PCR and site-directed mutagenesis permits the directed modification of genetic sequences, effectively reprogramming genes.

APPLICATION

The significant scientific developments described above set the stage for the biotechnology industry. Understanding the role of DNA in programming the abilities of individual cells, combined with knowledge of how information is encoded in DNA, the mechanisms by which cells use this information, and the development of molecular biology techniques to manipulate DNA, gave rise to modern biotechnology.

In 1973, Stanley Cohen at Stanford University and Herbert Boyer at the University of California at San Francisco developed methods to splice genes and express foreign proteins in bacteria. This made it possible to deliberately make defined changes to biological systems, permitting the directed modification of microbes and cell cultures to produce desired products. Boyer and venture capitalist Robert Swanson formed Genentech in 1976, a defining event in modern biotechnology. Genentech, one of the first biotechnology companies, aimed to commercialize gene splicing technology by initially producing recombinant human insulin in bacteria to treat diabetes.

Prior to 1976, drugs were either chemically synthesized or extracted from living sources. Before bacterial production, insulin was commonly extracted from pig pancreas and required the sacrifice of 50 animals to produce sufficient insulin for a single person for one year. The advent of gene splicing introduced new possibilities, facilitating drug development without screening libraries of chemicals and biological extracts, and enabled scientists to select proteins whose function was already known as lead compounds.

Following proof-of-principle production of a neurotransmitter, Genentech produced recombinant human insulin in bacteria in 1978, later to become the first recombinant DNA drug approved by the Food and Drug Administration.

In 1980, prior to FDA approval of its recombinant human insulin, Genentech capitalized on positive market sentiment towards biotechnology and raised $35 million in an initial public stock offering. Without the resources to fully develop and commercialize recombinant human insulin as a drug, Genentech had licensed manufacturing and distribution rights to Eli Lilly, the dominant supplier of beef and pig insulin. Aiming to independently develop and commercialize a drug, Genentech became the first biotechnology company to market its own biopharmaceutical product in 1985 when it used gene splicing to produce human growth hormone, a drug previously available only by harvesting pituitary glands from deceased human organ donors. Since then, Genentech has produced many additional products, was bought by Roche Pharmaceuticals, and was subsequently resold on the public markets.

Genentech focused on one of the first core technologies defining the biotechnology industry, but is not the first biotechnology company. That status belongs to Cetus. Cetus was founded in Berkeley, CA, in 1971 and initially focused on using automated methods to screen for microorganisms with industrial applications. Despite developing the Nobel Prize-winning polymerase chain reaction technology, the company was not able to maintain independence, and was acquired by Chiron in 1991.

COMMERCIALIZATION

The history of Genentech serves as a paradigm for biotechnology product development and corporate growth. Genentech was founded to exploit a novel scientific innovation. Without sufficient resources to fully develop and commercialize its first product, Genentech licensed these rights to a larger partner. Tapping revenues from early products enabled Genentech to develop sufficient bulk to fully research, develop, and commercialize its own products.

The means and motivation must exist in order to develop a biotechnology product. The motivating factor can be as simple as consumer demand, permitting a company to derive revenues from sales. Alternatively, if a technology is sufficiently appealing, the potential to create new markets can motivate development. Conversely, public resistance to biotechnology products, such as opposition to genetically modified crops, can exert a negative influence on the marketability of a product. Whether a company is compensated directly from sales, government

---

Box

Genentech: Commercializing a new technology

Genentech was founded in 1976 to capitalize on the revolutionary gene splicing technology developed by Stanley Cohen and Herbert Boyer. The company has since diversified to other technologies and boasts revenues in excess of $11 billion. It is also the only biotechnology company to never trade below its initial public offering (IPO) price, and has been profitable for all but two of its years as a public corporation.

1976: Genentech founded by Robert Swanson and Herbert Boyer

1977: Genentech produces the first human protein (somatostatin) in a microorganism

1978: Genentech produces human insulin in bacteria

1979: Genentech produces human growth hormone in bacteria

1980: Genentech goes public and sets a record for IPO stock price appreciation

1982: Genentech launches the first recombinant DNA drug: human insulin (licensed to Eli Lilly)

1985: Protropin (human growth hormone) approved by FDA— the first recombinant pharmaceutical product to be manufactured and marketed by a biotechnology company

1986: Roferon (interferon alpha-2a) approved by FDA and licensed to Hoffmann-La Roche

1987: Activase (tissue-plasminogen activator) approved

1990: Genentech's Hepatitis B vaccine, licensed to SmithKline Beecham, receives FDA approval

1993: Nutropin (somatropin) receives FDA approval
Pulmozyme receives FDA approval

1997: Rituxan receives FDA approval

1998: Herceptin (trastuzumab) receives FDA approval

1999: Roche exercises option to purchase ownership of Genentech, offers 22 million shares in an IPO six weeks later in the largest healthcare-related IPO ever
Secondary offering releases another 20 million shares, raising $2.87 billion in the largest secondary offering ever

2000: Third offering of 19 million shares raises $3.1 billion

grants, or awards, or is compensated indirectly from tax credits, there must be some motivation to support development.

Legal and regulatory pressures can promote or discourage development. Long development times and the relative ease of reverse-engineering necessitate intellectual property protection for biotechnology products. Patents grant the right to exclude others from practicing an invention, providing an incentive for patent holders or licensees to develop patented applications by preventing competitors from capitalizing on their research and development investments. For this reason, many biotechnology firms form around patented scientific methods or proprietary knowledge that create a barrier to competitors and a source of revenue through licensing of partially- or fully-developed products and technologies.

A characteristic distinguishing biotechnology (and pharmaceutical) products from those of many other industries is the requirement for rigorous and lengthy assessments to verify the safety and, in the case of drugs, efficacy, of products prior to being able to market them. Companies and financiers are therefore often unwilling to commit resources for development of drugs and other products for which the regulatory path is uncertain.

In addition to limiting development, government regulations can also motivate development. The Orphan Drug Act is an example of an incentive for drug development; tax credits and market exclusivity are granted to companies developing drugs for small populations that meet specific criteria.

Biotechnology development is fueled by innovation. The importance of specialized knowledge means that entrepreneurship by accomplished scientists is common in the genesis of biotechnology companies. The significant risk of product development failure compels biotechnology companies to focus on research and development until marketable products emerge. Patents and other barriers to entry are essential to prevent late-entering competitors from capitalizing on the efforts of pioneers.

INDUSTRY TRENDS

Many of the companies founded in the 1970s and 1980s sought to become fully vertically integrated drug developers, incorporating processes from drug discovery and development through production and sales. The prototypical company of this era aimed to develop treatments for unmet disease conditions and used the financing power of favorable public markets to fund expensive drug development efforts. Companies such as Genentech and Amgen were successful enough to achieve independence, but when market support for biotechnology disappeared, many companies had to reformulate their business models, merge, or liquidate.

Two impediments that prevented many of these early biotechnology companies from achieving vertical integration were the limited amount of available funding, which could not support the number of high-burn companies being founded, and the lack of experienced managers. The number of biotechnology companies aiming to become fully integrated diluted the amount of funding available at the time, limiting the support that each company could attain. Additionally, in order to develop vertically integrated companies, young startups needed managers with broad expertise from product development to commercialization. The only potential source for people with these skills was the pharmaceutical industry. Unfortunately, the pharmaceutical industry had divided the drug discovery and commercialization process into separate divisions managed by specialists, so no suitable managers existed. Furthermore, because biotechnology companies were seen as competitors, established pharmaceutical companies had little incentive for collaboration. By the late 1980s, pharmaceutical company sentiment towards biotechnology partnerships softened as pharmaceutical companies found themselves unable to maintain their growth rates solely by their internal research programs.

The 1990s saw the emergence of platform and tool-based companies seeking to commercialize drug targets, services, and technologies that could be sold or licensed to other companies.

Revenue streams emerged from partner licensing fees, royalties, and research contracts. Although revenues from tools and services can make a company profitable, there is always the risk that these offerings can become commodities or obsolete. Recognizing that revenues from tools and services could fund product development efforts, hybrid business models emerged in the late 1990s and early 2000s, capitalizing on the stability of tool and service sales while still selling the promise of product development. In addition to licensing or selling research tools to others, they were also used internally for product development. In principle, hybrid companies could therefore enjoy stable revenues from licensing and sales agreements while attracting investors by selling the promise of product development. The time and energy that must be devoted to marketing and selling tool offerings and keeping them current can make product development slower for hybrids than for product-focused companies. This reduced pace is balanced by the stability granted by revenues derived from tools which permit hybrid companies to better weather unfavorable financing environments.

The "no research, development only" (NRDO) model gained favor in the wake of the biotechnology bubble of 2000. A derivation of the specialty pharmaceutical model of seeking additional markets for drugs already approved in one or more countries, the goal of NRDO firms is to acquire promising lead compounds and manage their clinical trials, at which point the drugs can be marketed in partnership with, or sold to, larger firms. NRDO firms were able to capitalize on the wealth of drug leads and managers that could be inexpensively acquired from firms struggling or liquidating as a result of unfavorable market conditions. A limitation of the NRDO model derives from the reality that many important discoveries in science emerge in the course of unrelated research. By not participating directly in research, NRDO firms are unable to realize the significant upside of tangential discoveries that emerge from research. A lack of internal drug development talent also challenges managers to obtain skilled guidance, often from paid consultants or contract research laboratories rather than internal experts, to assess the quality of potential product acquisitions.

Another recent trend is the move toward larger-scale projects. The ability to automate procedures such as DNA sequencing, microarray analysis, and drug screening make it possible to perform research at an unprecedented scale. Data mining and massive bioinformatics projects have also formed the core of companies. This shift in scale demonstrates a very important change in the way research is conducted. The ability to perform large-scale experiments requires reliability and automation, attributes not often found in basic scientific discoveries and methods. DNA sequencing, a procedure that can now be fully automated, once required days of manual labor. Just as computers have advanced knowledge in other disciplines with their ability to process information and reliably and repeatedly perform tasks, the ability to automate biotechnology experiments will lead to greater discoveries at lower costs.

II

Science

Scientific research is a slow, painstaking process often fraught with setbacks; unfortunately, managers unfamiliar with this process fail to appreciate these difficulties. Because biotechnology involves novel products and techniques, it is difficult to predict the hurdles that will be encountered or the precise outcome of development efforts. Furthermore, a regulatory burden arises from the need to verify the safety and efficacy of biotechnology products. This section presents a detailed overview of relevant scientific topics to facilitate better understanding of challenges and opportunities of biotechnology research.

The biotechnology industry is not defined by a set of products, but by a set of enabling technologies. The prototypical biotechnology company focuses on research and development and uses molecular biology techniques to develop drugs and other useful products. Molecular biology is distinguished from general biology by the fundamental nature of the material studied. Whereas biology is the general study of life, molecular biology seeks to understand the inner workings of life's processes. Using molecular biology techniques, biotechnology companies are able to manipulate the fundamental processes responsible for diseases, or tap biology for other useful purposes.

Biotechnology companies engage in basic and applied research (see Figure 4.3). Basic research is primarily focused on acquiring new knowledge regarding the principles underlying phenomena and observations. Basic research is characterized by hypothesis testing, analytical experiments, and theory development. Building on basic research, applied research develops new knowledge and applications. Biotechnology firms use applied research to develop and commercialize the innovations and discoveries that emerge in the course of basic research.

Prior to the advent of molecular biology, biologists sought to answer such questions as how our physical characteristics are inherited from one generation to the next, how food is converted into energy, and how different cell types develop and perform their specialized roles. These researchers were able to identify agents responsible for disease and the role of human tissues in health and disease. It was not until the development of molecular biology that it became possible to discover and alter the actual processes responsible for health and disease states.

In 1953, Francis Crick and James Watson discovered the structure of DNA, the primary source of information in cells that permits genetic characteristics to be passed on from one generation to the next and bestows traits on cells. Following elucidation of the structure and function of DNA, the genetic code by which information is stored in genes was deciphered and the methods by which this information is ultimately translated were determined. These developments helped redefine biological research, but it took nearly 30 years—the first biotechnology drug was approved in 1982—for modern biotechnology to demonstrate its potential.

Understanding the fundamentals of molecular biology, combined with the ability to introduce genes into organisms, enabled biotechnology: the directed modification of living things toward useful ends.

As the science has matured, companies in previously unrelated industries have invested increasingly in biotechnology research. The application of biotechnology in diverse industries makes it difficult to define biotechnology companies discretely; every company that uses biotechnology is not a biotechnology company. Instead of being defined solely by their research activities, biotechnology companies are defined by the concentration of their focus on biotechnology research and development.

The application of biotechnology provides new answers to old problems, but also introduces new challenges. Markets for many applications are well established. The question in these cases is not whether the customers exist, but if it is possible to produce a useful and compelling product at a reasonable cost in a reasonable amount of time.

Chapter 3

Introduction to Molecular Biology

Everything should be as simple as possible, but not simpler.
Albert Einstein

Biotechnology research seeks to develop applications of molecular biology. Many sources use analogies to recipe books or blueprints to explain the role of DNA and genes in molecular biology. Ultimately, these analogies obscure the importance of topics such as regulation of gene expression, which is of fundamental importance in understanding molecular biology. When applying one's knowledge of biotechnology fundamentals, most metaphors fail. It is only by understanding molecular biology and biotechnology applications that one can appreciate the applications and limitations of techniques used in molecular biology.

This chapter presents a brief, metaphor-free, introduction to molecular biology. Subsequent chapters describe the tools, techniques, and applications of biotechnology and provide greater details on the potential and limitations of molecular biology.

INFORMATION FLOW IN MOLECULAR BIOLOGY

In order to understand the basis of most biotechnology applications, it is necessary to first understand the process by which information in genes leads to the formation of structural and functional proteins.

transcription   *translation*

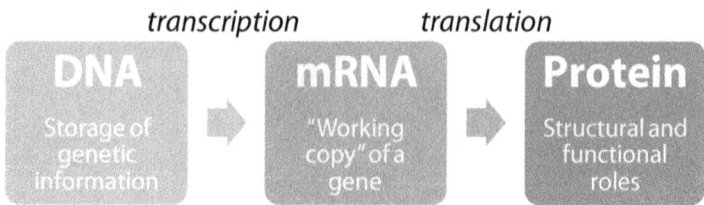

DNA ➡ **mRNA** ➡ **Protein**

| Storage of genetic information | "Working copy" of a gene | Structural and functional roles |

Figure 3.1 *Simplified model of information flow in molecular biology*

Proteins serve structural and functional roles that give individual cells—and by extension whole organisms—specific structures and functional characteristics. When many people think of proteins, they think of foods such as meat and beans. While animal muscle and plant seeds are excellent sources of dietary protein, proteins play a central role in all cell types and perform functional and structural roles (see Table 3.1). Examples of structural proteins include keratin, which makes skin waterproof, and myosin, which interacts with other proteins in muscles to make them flex.

DNA contains information that describes the construction of proteins. The process of protein synthesis is as follows:

1. DNA contains the information to produce proteins.
2. Information encoded in DNA is *transcribed* into a molecule called messenger RNA (mRNA)—effectively a "working copy" of the DNA sequence of a given gene.
3. mRNA is *translated* into proteins by the protein synthesis machinery, the composition of the resulting protein corresponding to the original DNA instructions.

This basic mechanism is conserved in all life forms, from bacteria to humans. The implication of this common process that converts information in DNA into functional proteins is that similar techniques can be used to investigate and manipulate all biological systems. Furthermore, it is possible to make human therapeutic proteins, for example, in organisms as distantly related as bacteria.

Understanding the roles of DNA, RNA, and protein and their relationships to each other is essential to understanding molecular biology. While there are some specific exceptions (e.g., retroviruses and prions) to the order and direction of information flow shown in Figure 3.1, these examples still fit within the general framework, and the majority of biological systems use the framework as presented.

DNA: STORING AND RELAYING INFORMATION

Deoxyribonucleic acid (DNA) is the primary source of genetic information in cells. Humans, plants, animals, and bacteria all contain DNA. DNA is physically passed from generation to generation, bestowing certain traits of parents to their children. The reason why children have physical characteristics from each of their parents—a child may have their mother's eye color and father's hair color—is because they received half their

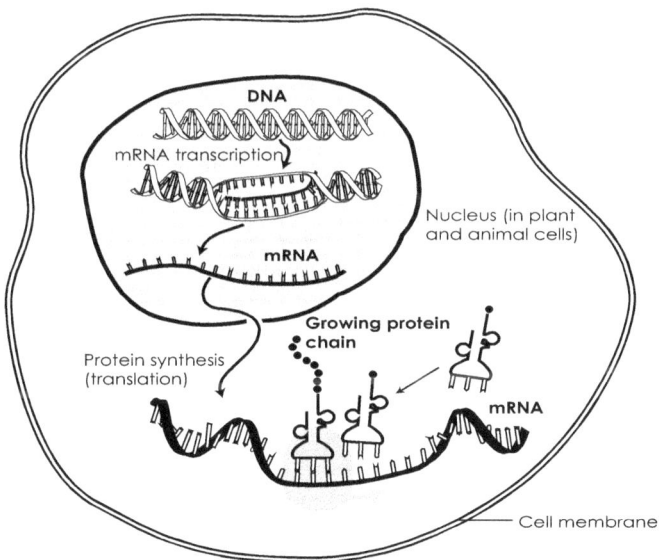

Figure 3.2 *General scheme of gene expression*
Modified from National Human Genome Research Institute

DNA from each parent. Each of our cells (with a few exceptions like red blood cells, eggs, and sperm) contain all the DNA required to code our genetic features. Individual regions of DNA that confer traits are called genes. Information in genes is relayed to the protein synthesis machinery within cells where it dictates the production of proteins. The word "genome" refers to all the DNA in an organism. The human genome contains approximately 30,000 genes arrayed on 46 long stretches of DNA called chromosomes.

DNA is essentially composed of two intertwined strands that form a double helix. The two strands of DNA are said to be complementary because the sequence of one strand indicates the sequence of the opposite strand, like a photograph and its negative. Each strand is physically composed of four different chemical units called nucleotides, the sequence of which encodes the genetic information. These four chemical units, adenine, cytosine, guanine, and thymine, are often abbreviated as A, C, G, and T, respectively. Just as the English language can be expressed in twenty-six letters, the genetic code is expressed in these four chemical units. A DNA "sequence" refers to the specific order of A's, C's, G's, and T's in a stretch of DNA.

There are two essential elements of genes: coding and regulatory elements. The coding elements of genes are first transcribed as mRNA, which is then translated into protein. The chemical sequence of A's, C's, G's, and T's in the coding region of a gene determines the composition and structure of the resulting protein and, by extension, its function. Regulatory elements affect the rate at which genes are transcribed and translated, and may be interspersed within the coding sequence or outside of it. Regulatory elements also control the cell types within which specific genes are activated, and the timing and magnitude of gene expression. Gene regulation thereby allows individual proteins to be expressed only in certain cells at specific times and at specific rates.

Proper regulation of gene expression—the production of gene products—is essential. Under- or over-expression of genes

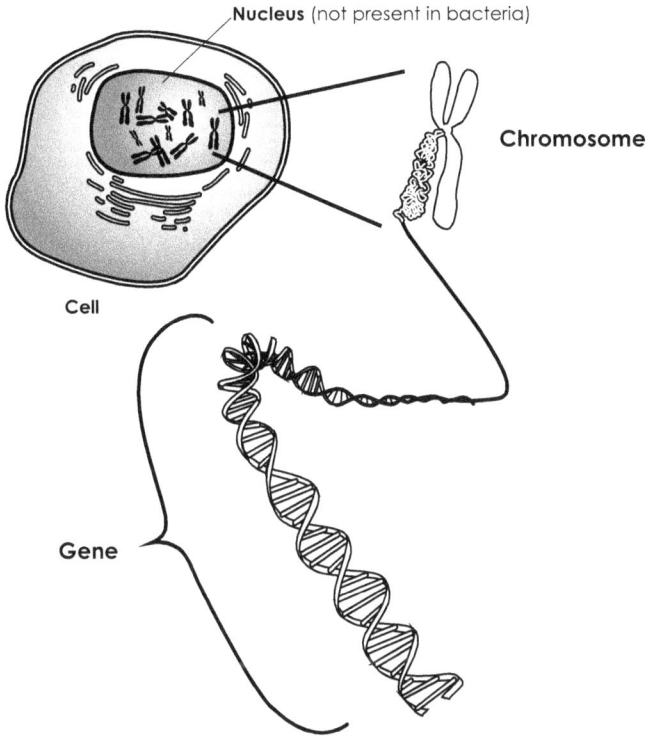

Figure 3.3 *DNA: Chromosomes and genes*
Modified from National Human Genome Research Institute

can have deleterious effects. For example, many forms of cancer are caused by mis-regulation of gene expression that results in uncontrolled cell division. A potential solution for diseases resulting from low expression of genes is to use gene therapy to introduce affected genes or regulatory elements to spur additional production. One of the challenges of gene therapy is developing methods to regulate the expression of genes that are introduced into cells and ensure that they are not over-expressed. A solution for diseases caused by over-expressed genes is RNA interference. This procedure prevents translation of mRNA, inhibiting protein production. RNA interference is described in further detail in Chapter 6.

Human chromosomes and genetic trait inheritance

The human genome is composed of chromosomes. We get 23 chromosomes from our mother and 23 chromosomes from our father, constituting 23 pairs. While 22 of the 23 chromosome pairs are similar in both men and women, the 23rd pair is quite different and determines the sex of an individual. For the 23rd pair of chromosomes, women have two X-chromosomes while men have one X- and one Y-chromosome. Because X-chromosomes contain more DNA than Y-chromosomes, they are physically larger than Y-chromosomes. Having too many or too few chromosomes can affect gene regulation and cause diseases. Down's Syndrome, for example, occurs in individuals with three copies of chromosome 21.

The roles of X- and Y-chromosomes are important in understanding sex-linked diseases. Women do not have Y-chromosomes, so diseases that are caused by defective genes on the Y-chromosome can only occur in men. Additionally, men only have one X-chromosome, so mutations in genes on the X-chromosome are more likely to affect males, because the second X-chromosome in women can sometimes compensate for mutations on the first. Color blindness, caused by a mutation on the X-chromosome, is more common in men than women for this reason.

mRNA: THE MESSENGER

Messenger RNA (mRNA) is used to relay information from genes in DNA to the protein synthesis machinery. An additional feature of mRNA is that it can be destroyed once sufficient protein is produced, permitting an extra level of control of gene expression. RNA is also present in forms other than mRNA, some of which are described later in this chapter.

It is possible to affect expression of genes by targeting their mRNA with antisense RNA or DNA—nucleic acids which can bind the mRNA. Unlike DNA, which is usually double-stranded, mRNA is single stranded. Nucleic acids (DNA or RNA) containing a sequence that can bind to a given mRNA will prevent translation by the protein synthesis machinery, inhibiting gene expression. The Flavr Savr tomato, a tomato engineered to have a long shelf life, was produced by introducing antisense RNA corresponding to mRNA for an enzyme involved in fruit spoilage. Inhibiting expression of this gene delays spoilage. In 1998 the FDA approved Isis Pharmaceuticals' Vitravene, the first antisense drug, to treat cytomegalovirus-induced retinitis.

TRANSLATION: MAKING PROTEINS

Just as DNA and RNA are composed of linked nucleotides, proteins are comprised of chains of amino acid units. When mRNA is translated to produce a protein, the protein-synthesis machinery "reads" the nucleotides three at a time, assembling amino acid chains that correspond to the mRNA sequence. The basic elements of the protein-synthesis machinery are tRNA, a form of RNA that *transfers* amino acids to the protein-synthesis machinery in a way that enables them to be linked together, and ribosomes, which help form the chemical bonds that attach amino acids in a protein chain.

The three-nucleotide sequence elements on mRNA that code for individual amino acids are called codons. These are

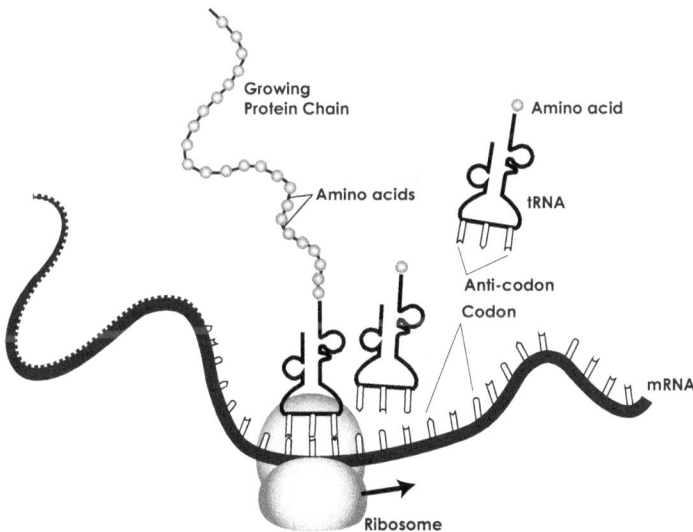

Figure 3.4 *Protein translation*
Modified from National Human Genome Research Institute

matched by anti-codons on tRNA to ensure that the appropriate amino acid is aligned with a given mRNA sequence. The 64 possible combinations of A, C, G, and T at each codon code for only 20 different amino acids. This redundancy in the genetic code, permitting multiple codons to specify common amino acids, is considered a form of protection against DNA mutations and has applications in identifying foreign DNA from sources such as viruses which may use different "dialects" of the genetic code.

The chemical characteristics of amino acids in a protein cause it to fold into a defined 3-dimensional structure. That determines the protein's function. Because the DNA sequence of a gene dictates the sequence of amino acids in a protein, and the sequence of these acids in a protein determines its structure, one can deduce a protein sequence, and potentially its structure and function, from the gene sequence encoding it.

PROTEINS AND ENZYMES

Proteins, the workhorses of cells, are responsible for the majority of structural features and functional characteristics in cells. Enzymes are proteins that perform functional roles as part of the cellular process. Different types of cells get their characteristics by expressing a specific array of genes, resulting in production of a complement of proteins that give each cell type its unique characteristics. Pancreatic cells, for example, produce the protein insulin to regulate blood sugar levels; neurons produce neurotransmitters essential for brain function; and hemoglobin is made in blood cells, enabling them to carry oxygen. Examples of enzymes include proteases that break down proteins or enable digestion of food, and polymerases that assemble DNA and RNA. Some genes are expressed only in certain cell types whereas others are widely expressed. Examples of widely-expressed genes include those encoding proteins and enzymes involved in general cellular activities such as DNA replication, mRNA translation, protein synthesis, energy production and maintenance of structural integrity.

Table 3.1 *Examples of protein and enzyme functions*

Enzyme	Function
Amylase	Breaks down starches and other complex carbohydrates into basic sugars
Cellulase	Breaks down cellulose, found in the cell walls of plants
Lipase	Breaks down fats
Protease	Breaks down proteins

Protein	Function
Collagen	Main protein in connective tissue; structural roles in skin, cartilage, teeth, bone, and other tissues
Keratin	Makes skin waterproof and contributes to strength and flexibility
Myosin	Muscle contraction

Production of inappropriate proteins in cell types and mis-regulation of protein expression are at the root of many diseases. As mentioned above, many cancers result from mis-regulation of gene expression that causes uncontrolled cell division.

Molecular biologists can transfer genes from humans and other animals into bacteria, yeast, and other organisms to confer the ability to produce specific proteins that may be extracted for therapeutic use. For example, Genentech produced its first drug by introducing the gene for human insulin into bacteria and extracted the resulting protein to produce a treatment for human diabetes. Genes can also be transferred from one organism to another to confer new attributes. Pesticide-resistant crops have been produced by incorporating naturally-occuring pesticidal proteins into plants. Bacteria have also been modified to perform roles such as decomposing oil spills by adding genes encoding proteins with the ability to break down components of oil. Additional examples are described in Chapter 6.

OTHER FORMS OF RNA

Traditional molecular biology held that the primary role of RNA in cells was largely limited to housekeeping functions such as transferring information from DNA to the protein syn-

Table 3.2 Selected *RNA types*

RNA type	Function
mRNA	Messenger RNA. Contains a working copy of a gene sequence and is read by the protein synthesis machinery to produce proteins.
tRNA	Transfer RNA. Transfers amino acids to the protein synthesis machinery to produce proteins.
rRNA	Ribosomal RNA. Part of the protein synthesis machinery. Also useful for determining evolutionary similarity between organisms.
aRNA	Antisense RNA. Used for gene regulation.
siRNA	Small Interfering RNA. Used for gene regulation.
snRNA	Small Nuclear RNA. Used to edit mRNA, regulate gene expression, and maintain chromosome tips (telomeres).

thesis machinery (mRNA), transporting amino acids to be assembled into proteins (tRNA), and translating mRNA into protein (rRNA).

Sidney Altman and Thomas Cech shared the 1989 Nobel Prize in Chemistry for their discovery of catalytic properties of RNA. The ability to catalyze (increase the rate of) biochemical reactions had previously been thought to only exist in proteins. Altman and Cech found a role for RNA in the splicing of mRNAs, ultimately making it possible for a single gene to give rise to several different proteins. The significance of Altman and Cech's discovery was expanded more than a decade after they received the Nobel Prize. Following sequencing of the human genome it was discovered that the human genome contained only a fraction of the genes previously thought necessary to produce the complete set of proteins comprising human biology. The ability of this small set of genes to produce the full complement of human proteins could largely be explained through mRNA splicing.

More recently myriad forms of RNA have been discovered, and diverse roles for RNA have also been elucidated (see Table 3.2). These discoveries indicate that controlling cellular activities is more complex than previously thought, suggesting that there are also more opportunities to influence cellular activities.

THE BIG PICTURE

Genes interact with the environment and with each other to confer traits. While the presence or absence of a gene can potentially confer a given trait, environmental factors also play a role. Our physical characteristics are a combination of genetic and environmental factors. A child with a hypothetical *tallness* gene, for instance, would not necessarily grow taller than a child without the gene; the child with the *tallness* gene would also require adequate nutrition to fuel the extra growth (and the effect of the *tallness* gene may be limited or enhanced by the action of other genes). Rather than thinking of genes as determinants of physical characteristics, they should be regarded as potentials or predispositions for characteristics.

The ability to modify characteristics of cells is similarly limited by biological and physical constraints. Since some cells are rapidly replaced, induced changes will be quickly lost. Other cells are dormant, precluding their potential to express modifications.

Furthermore, biology is complicated. In fields such as industrial chemistry or engineering, applications are developed from well-characterized principles. With biotechnology on the leading edge of molecular biology research, it can be difficult or impossible to foretell the outcomes of manipulations and they can have unforeseen consequences. Because it is not possible to fully predict the outcome of these procedures, scientists must perform experiments, take observations, refine theories, and finally develop functional applications. This is why biotechnology research is so complex, time consuming, and fraught with unforeseen setbacks and disappointments.

Chapter 4

Drug Development

It is only by the means of the sciences of life that the
quality of life can be radically changed.
Aldous Huxley

Drugs are substances that affect the functions of living
things and are administered to treat, prevent, or cure
unwanted diseases and symptoms. The United States
Food and Drug Administration (FDA) regulates drug marketing, requiring manufacturers to prove their products to be
safe, effective, and appropriately labeled. The drug development
process identifies drug candidates and subjects them to increasingly stringent tests to assess their safety and efficacy. Drug
development is paradigmatic of the general process by which
biotechnology products are developed, with one important difference: non-therapeutic products are not subject to the same
regulatory pressures to gain marketing approval.

Scientists start with simple, defined, model systems that enable them to identify potential drugs. These potential drugs are
then tested in increasingly complex and real-world situations
to prove their efficacy. It is important to test for as many con-
tingencies as possible. Something that works well in a simple
model system may fail in real-world use due to any number of
unforeseen circumstances.

The description of drug development in this section is presented as a model for biotechnology product development. Producing and selling drugs consists of three basic stages: discovery, development, and commercialization. Less than 1 percent

of early candidate compounds make it through the drug development process.

Discovery-stage research produces lead compounds that must pass tests to predict their toxicity, to determine if they can be effectively administered, and to project the likelihood of recouping development costs (see Figure 4.4).

The development process builds on observations and products from discovery-stage research. Formulations are developed to optimize drug administration, and pre-clinical and clinical trials are employed to test the safety and efficacy of drugs in humans. In addition to testing the physical properties of a drug, manufacturing processes which can consistently produce doses of equivalent purity and efficacy over a period of time must be developed and tested.

The products of development-stage activities—approved drugs—are commercialized. Drug marketing, one of the dominant elements of commercialization, is described in Chapter 12.

BIOTECHNOLOGY VS. PHARMACEUTICAL DRUG DEVELOPMENT

Traditional pharmaceutical drugs differ from biotechnology-derived drugs in the methods by which they are discovered and manufactured. As a result, the resulting drugs have markedly different characteristics.

To distinguish traditional pharmaceutical from biotechnology drug development, consider the traditional pharmaceutical and biotechnology forms of therapeutic insulin. Prior to Genentech's production of recombinant human insulin, pharmaceutical companies extracted insulin from the pancreas of pigs, cows, and horses. Glands from fifty pigs were needed to produce sufficient insulin to treat a single person for one year. Insulin from these sources was subject to disease transmission, shortages, and reactions with the human immune system. Genentech produced recombinant human insulin, the first biotechnology drug, by synthesizing it in bacteria. Bacterial fer-

Small-Molecule Drug

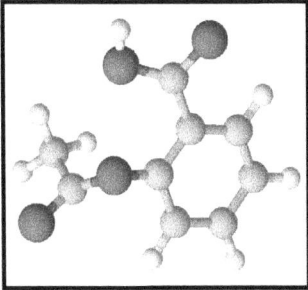

Biologic Drug

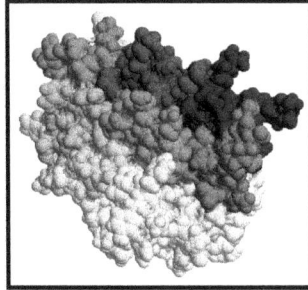

Aspirin
23 atoms

Erythropoietin
1297 atoms

Figure 4.1 *Small-molecule and biologic drugs*

mentation allowed for greater production capacity, avoidance of immune system reactions typical of non-human forms of insulin, and elimination of the threat of transmission of animal diseases. This example illustrates how the fundamentals-based approach to product development employed by biotechnology firms permits the development of solutions not attainable by traditional pharmaceutical development.

Traditional pharmaceutical drug discovery was based on trial-and-error screening of synthetic compounds and directed selection of biological extracts that can affect model systems. The emphasis of research was to understand biological systems in order to find potential drug targets. Compounds and extracts that interacted with these targets were then selected for further study to see if they could be used as drugs.

The molecular biology techniques used by biotechnology firms differ from traditional pharmaceutical development because they permit a finer-scale analysis of biological systems and the directed design of biological compounds as drug candidates. Traditional pharmaceutical development was limited to chemical synthesis and biological extracts. The reason for this limitation was that traditional pharmaceutical development originated before the advent of molecular biology techniques,

2006 Biotechnology drug sales by category

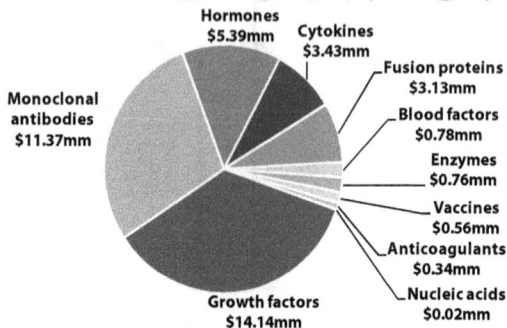

Figure 4.2 *Biotechnology drug categories*

which enable the directed design and production of biological molecules.

Drugs produced by traditional pharmaceutical means tend to be small molecules that are orally doseable as tablets, capsules, or liquids (see Figure 4.1). Following absorption in the gastrointestinal tract these drugs travel throughout the body in the bloodstream, and can often be mass-produced for a relatively low cost.

While biotechnology research techniques do enable new possibilities, they have not rendered pharmaceutical research techniques obsolete. Traditional pharmaceutical research is still practised because of available expertise, the abundance of chemical and biological-extract libraries, and the strength of techniques for target selection and optimization.

The majority of biotechnology drugs have been proteins, such as growth factors, monoclonal antibodies, hormones, and cytokines (see Figure 4.2). Other categories include nucleic acids and vaccines. The ability to design, modify, and synthesize biological compounds means that many biotechnology drugs are larger and more complex than traditional pharmaceutical drugs.

Drug delivery is an issue for biotechnology-derived drugs because proteins and other biotechnology drugs such as nucle-

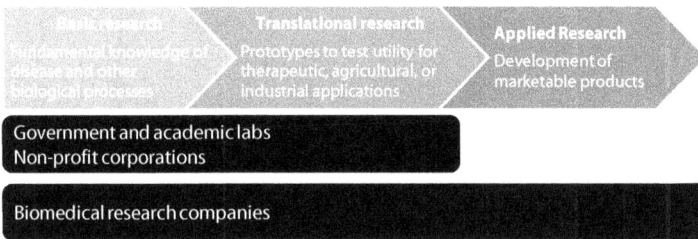

Figure 4.3 *Basic and applied research*

ic acids are less likely to survive the acidic conditions in the stomach and are generally unable to pass through the intestinal lining and travel the bloodstream to their therapeutic target. Biotechnology drug delivery techniques include injection, skin patches, and inhalation. Some of these alternative delivery systems allow for more precise tissue targeting and improved dosage control, presenting new opportunities.

Because the post-discovery activities for biotechnology-derived and traditional pharmaceutical drugs are relatively similar, modern pharmaceutical firms are also able to develop lead compounds generated by biotechnology firms, and actively engage in biotechnology research themselves. Conversely, as the biotechnology industry has matured, some biotechnology companies have developed late-stage development and marketing abilities on par with small and medium-sized pharmaceutical companies. These shifts in the roles of pharmaceutical and biotechnology firms have led to a blurring of the distinction between the two.

THE FIVE BASIC STEPS OF DRUG DEVELOPMENT

There are two fundamentally different types of research conducted in early-stage drug development: basic and applied research (see Figure 4.3). Basic research is directed at improving fundamental knowledge of biological systems, disease processes, and potential points of therapeutic intervention. Applied research utilizes this knowledge to identify and develop

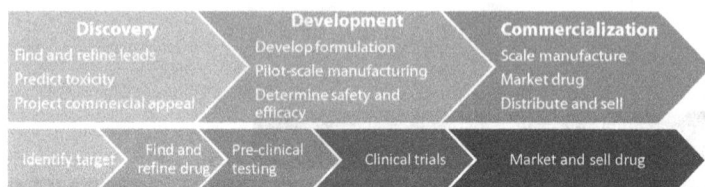

Figure 4.4 *The process of drug development*

the therapeutic agents themselves. This distinction is important because different toolsets and mindsets are involved in these two types of research. Basic research generally does not directly produce drugs. Instead, it lays the foundations upon which drugs are developed and produced. Genomics, proteomics, molecular physiology, and other basic research areas contribute essential information for disease characterization and target identification. Applied research is required to enable further development. Translational research forms a bridge between basic and applied research, testing principles from basic research and generating "proof-of-principle" in preparation for applied research.

STEP 1: IDENTIFY A USEFUL DISEASE TARGET

A fundamental understanding of the science involved in a therapeutic problem is essential for drug development. Ideally, scientists start with an understanding of the molecular processes affecting conditions they wish to treat and test hypotheses about which drugs are likely to be effective for a condition.

Model systems are employed because it is unethical and impractical to test uncharacterized candidate drugs on humans. These model systems generally progress in complexity through the stages of drug development. Early-stage model systems may be as simple as a set of molecules in a test tube (subject to knowledge of disease processes) and later-stage model systems may be live animals with diseases similar to human conditions (subject to availability of relevant disease models).

To work on a problem effectively, the tools must be well-defined. Poor model systems or a lack of precise measures can yield experimental results which are misleading or difficult to interpret. It is vitally important to have a fundamental appreciation of the problem to be solved and to apply the appropriate tools to assess potential solutions. Many biotechnology applications, such as gene therapy and RNA interference (described in Chapter 6) are challenged by poor availability of predictive models.

STEP 2: FIND AND REFINE A LEAD COMPOUND

Potential lead compounds typically originate from one of two sources: purified naturally occurring compounds or *de novo* design and synthesis of new compounds. As described earlier in this chapter, biotechnology drugs have different properties and capabilities than traditional pharmaceutical drugs. The traditional pharmaceutical method for drug discovery involves screening libraries of natural or synthetic compounds to find those that achieve the desired effect in model systems. The molecular biology techniques used in biotechnology drug development permit directed selection of natural compounds—usually biological molecules—and the design and synthesis of novel biological compounds as drug candidates.

As knowledge of biological systems improves, it becomes increasingly possible to refine screening methods. Knowledge of disease mechanisms can also help build better model systems through identification of appropriate therapeutic targets as well as other targets which might be a source of undesirable side effects.

An understanding of how biological systems operate at the molecular level, combined with the ability to express gene products in bacteria, yeast, and animal-derived cells, enables development of drugs based on specific disease requirements. Many traditional pharmaceutical drugs were discovered by screening libraries for effective leads, rather than starting with knowledge disease processes and working backwards to find a solution. Knowledge of the structure of a key molecule involved in a disease, such as the HIV protease that is integral to AIDS, enables *in silico* (computer model-based) techniques, using computers to select or design compounds likely to inhibit the enzyme.

Studying herbal remedies used by different cultures, indigenous peoples, and animals is also an excellent source for potential drug compounds. Some naturally occurring compounds, such as penicillin, are used as drugs based on their natural activities in biological systems. The antibiotic penicillin was identified as the factor that permits *Penicillium* mold to inhibit bacterial growth. Whereas penicillin is naturally produced by *Penicillium* as an antibiotic, the same therapeutic application it is used for in humans, other natural compounds, such as *botulinum* toxin, are used for novel purposes such as treating abnormal muscle contractions and in cosmetic applications.

Another method to discover new drugs is to examine the side effects of existing drugs. Minoxidil, now prescribed as a topical treatment for hair loss, was initially intended for the treatment of severe blood pressure. The curious side effect of stimulating hair growth led to its use as a treatment for balding. In another example, Viagra's potential for the treatment of erectile dysfunction was discovered in clinical trials for treatment of angina.

Once a potential drug that works in a model system is identified, it is time to study and refine its activity. This potential drug is called a lead compound. Aside from drug activity, factors such as a drug's shelf life at different temperatures, ease of large-scale manufacture, and lot-to-lot consistency must also be considered. In optimizing lead compounds, researchers aim to identify the elements that are essential for their activity and modify those elements to obtain optimal efficacy and/or safety properties.

STEP 3: TEST LEAD IN PRE-CLINICAL DEVELOPMENT

In pre-clinical development, lead compounds that emerge from the lead optimization process are subjected to a range of standardized animal, cellular, and biochemical tests designed

Drug development stages

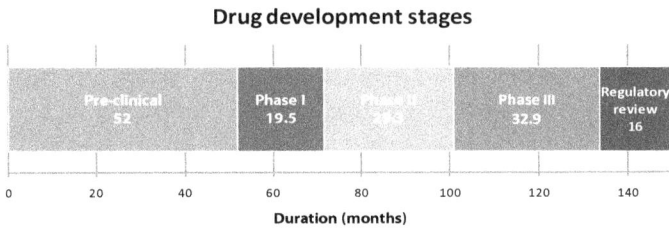

Figure 4.5 *Biotechnology drug development time*
Source: Dimasi, J.A., Grabowski, H.G. The cost of biopharmaceutical
   R&D: Is biotech different? *Managerial and decision economics*, 2007.
   28:469-479.

to gauge their suitability and safety for human administration as well as to estimate the range of dose levels of the compound that will be utilized in subsequent human trials. Animal models are also used to provide preliminary assessments of the absorption, degradation, and potential toxicity of drugs. During pre-clinical development, these animal tests are conducted under much more tightly prescribed conditions, such as the industry-standard GLP (Good Laboratory Practices) procedures, which have stringent quality control and quality assurance oversight than is normally used in earlier stages. While it may be possible to produce a desired effect in a model system in a laboratory setting, real-world situations often present unforeseen obstacles. For example, many gene therapy techniques that work in cultured cells in laboratory settings fail when introduced into human beings.

Scientists use animals to test toxicity and attempt to cure animal versions of human diseases before proceeding to human trials. Because biotechnology enables biological changes that were previously impossible, it is not possible to predict all the implications. Many products that work well in laboratory tests fail in clinical settings for unforeseen or even improbable reasons; drugs may not be taken up properly by cells; they may be metabolized into inactive or toxic forms by the liver; they may interact with other parts of the body to produce undesired effects; or they may simply not be sufficiently active. This is one of the reasons why animals are so important in drug research.

While success in animal tests does not necessarily mean that a compound will work in humans, a compound that performs poorly in animals is unlikely to work well in humans. Ultimately, human testing is necessary for the safety and efficacy determinations required for FDA approval.

In addition to testing the drug candidate in animal models, another set of activities that takes place during pre-clinical development is development of methods for manufacturing and formulating the drug on a commercial scale. Unlike research costs, manufacturing costs recur over the life of a drug. Minimizing these recurring costs can significantly impact profits. This examination is also important for patent protection because it may lead to additional patent claims, potentially impeding the development of competing drugs.

STEP 4: CLINICAL TRIALS IN HUMANS

The clinical trial process, described in Chapter 8, investigates drugs for safety and efficacy in humans. There are four "phases" of clinical trials. Phases I through III demonstrate safety and efficacy prior to approval, and Phase IV monitors safety post-approval and tests new treatment indications.

Briefly, drugs are first tested in a small group of healthy, or in some cases affected, individuals to determine safe dosage limits. Larger trials follow to investigate safety in diverse populations, establish dosage regimens, and demonstrate efficacy. Drugs administered in clinical trials must be produced using current good manufacturing practices (cGMP), ensuring proper control of facilities, raw materials handling, manufacturing, and associated documentation. Clinical trial data is submitted to the FDA as part of a New Drug Application (NDA) or Biologics License Application (BLA).

STEP 5: OBTAIN APPROVAL; MARKET AND SELL DRUG

The FDA requires that drugs be approved prior to marketing. While safety is the primary concern, a drug with detrimental side effects may be acceptable if there are no better treatments and the severity of the disease warrants it.

Current estimates of development times for small-molecule drugs are 10-15 years with an estimated average cost of $802 million per approved drug.[1] It is worth noting that half of this cost is attributed to financing costs, reflecting the "opportunity-cost of capital" invested over the 10-15 year timeline. Roughly one-third of the expenditures are attributed to pre-clinical activities and the remaining two-thirds to clinical activities (for further discussion of the cost of drug development, see Box *The cost of drug development* in Chapter 11). The Tufts Center for the Study of Drug Development found that only five in five thousand small-molecule compounds that enter pre-clinical testing make it to human testing. Of these five, only one is approved.[2] These numbers were derived from examination of small molecule synthetic drugs, which are produced by traditional pharmaceutical techniques and differ from biologic drugs in several important ways. Estimates for the cost of biologic drug development are $1.2 billion, comprised of approximately $500 million in out-of-pocket expenses, and $700 million in capitalization costs.[3] Both these cost estimates include the cost of failed leads. Therefore, they do not predict the expenditures required to produce a single drug; they predict the investments required by successful and failed research projects that result in the development of a single drug.

1 DiMasi, J.A., Hansen, R.W., Grabowski, H.G. The price of innovation: new estimates of drug development costs. *Journal of Health Economics*, 2003. 22:151–185.

2 How new drugs move through the development and approval process. *Tufts Center for the Study of Drug Development*, November 1, 2001.

3 Kaitin, K.I. (ed.) Cost to develop new biotech products is estimated to average $1.2 billion. *Tufts Center for the Study of Drug Development Impact Report*, 2006. Vol. 8.

Relative to small molecule drugs, the sample size to evaluate development times and costs for biologics is much smaller and subject to bias—the first biologics had shorter development times than later entrants—but initial indications are that development times and costs for biologics are similar to those for small molecules. It is important to note that any estimate of drug development time or cost is profoundly affected by context. First-in-class drugs, drugs serving new markets, or drugs serving pressing needs are likely to require smaller and fewer clinical trials than drugs with little differentiation from existing alternatives or those serving less-pressing needs, decreasing the time and cost of development.

Once a drug receives regulatory clearance for marketing, it will likely be protected by patents that were filed before clinical trials began. With an average of twelve years of patent protection remaining after FDA approval, marketing and sales efforts must generate revenues and expand market penetration to

---

Box

The Human Genome Project and drug development

The Human Genome Project (HGP) was a multi-billion dollar multinational effort to sequence the entirety of the human genome, identify all the genes, improve tools for genome analysis, and address related ethical, legal, and social issues. The project started in 1990 and sequencing was completed ahead of schedule in 2003.

The initial findings from the HGP informed scientists that molecular biology was far more complicated than previously believed. Projections for the number of genes in the genome, for example, ranged from the high tens-of-thousands to more than 100,000. After examining the sequence of the human genome, it was found that the genome contained far fewer genes than previously suspected— less than 30,000. The key to enabling the genome to produce a sufficient diversity of proteins from this relatively small set of genes is in editing individual gene mRNAs so that each gene can produce multiple proteins. Entirely new methods for regulation of gene expression were also discovered, further complicating efforts to tame molecular biology.

So, why has the HGP not yet produced a revolution in drug development?

- The HGP was primarily a basic science endeavor. It has greatly expanded the knowledge-base essential for drug development
- Information gleaned from the HGP must be interpreted and understood before it can produce new drug leads
- It can take more than a decade for a drug lead to gain regulatory approval
- Most drug leads fail to gain regulatory approval

While processing the new information from the HGP will occupy researchers for decades to come, there are some immediate benefits. The project brought many technological advances. The costs of synthesizing and sequencing DNA, for example, decreased by several orders of magnitude and the precision of many experiments has improved, along with the ability to automate many procedures. These improvements have translated beyond humans into agriculture, where farmers are better equipped to identify top-performing plants and select them for traditional breeding programs. DNA fingerprinting, which has greatly advanced the field of forensics, is also a spin-off of the human genome project. These advances and others are benefitting science today, while we await the other outcomes of the HGP.

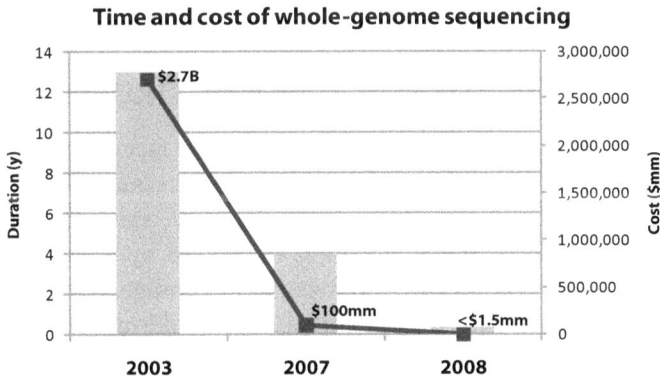

Figure 4.6 *Declining time and cost of human genome sequencing*
Source: James Watson's genome sequenced at high speed. *Nature*, 2008. 452:788.

deliver a return on R&D expenditures. After a drug is on the market, drug sponsors must monitor patients for unexpected side effects. Independent or sponsored clinical trials can also test suitability for additional indications, potentially expanding the market. The emergence of competing products and looming patent expiration dates motivate the development of alternative drug forms and formulations to leverage established brands, and modification of marketing methods to extend sales. These strategies are covered in greater detail in Chapter 11. Some firms specialize in modifying patented drugs, capitalizing on the negative specter of patent expirations, and patent their modifications to license them to pioneers as a means to preempt generics.

The aforementioned five steps of drug development must always occur, and the influences of individual biotechnology innovations are compartmentalized. Each innovation can only affect one or a few steps. For example, functional genomics can aid in developing model systems and selecting or designing potential drugs, but cannot resolve drug delivery or manufacturing issues. Molecular evolution can refine lead compounds, but cannot assist discovery, clinical trials, or manufacturing. The impact of this compartmentalization is that no single technology can profoundly alter the process of drug development; a number of complementary innovations are necessary for a revolution.

QUALITIES OF A "GOOD" DRUG

Beyond safety and efficacy, many other factors influence the quality of a drug. An ideal drug must address a market that is willing to pay a price that permits profitable sales. The commercial attractiveness of a drug is influenced by the size of a drug's patient population, the frequency of dosage, cost of production, barriers to entry of competitors, and availability and cost of alternative treatments.

Consider the example of penicillin, the first antibiotic. The devastating effects of bacterial infections and the absence of effective treatments guaranteed strong sales. The raw materials

Table 4.1 *Qualities of a "good" drug*

Market quality	Pharmacological qualities	Barriers to entry
• Market size • Dosage frequency • Manufacturing cost • Price elasticity	• Safety • Effectiveness • Chemical stability • Metabolic stability • Drug delivery	• Intellectual property protection • Market exclusivity • Challenge of generic manufacture

for penicillin production were known and initial tests showed a good safety profile. The remaining challenge for penicillin production was developing a method to produce sufficient quantities of the drug at a reasonable price.

In addition to safety and efficacy considerations, practical aspects such as chemical stability and therapeutic administration also affect the commercial prospects of a drug. Oral administration and patches are generally preferred to injections. Unfavorable administration can reduce patient compliance or place a drug at a competitive disadvantage to alternatives. Some drugs must also be mixed with carriers, which can impart their own side effects (see *Drug Delivery* in Chapter 5).

The chemical stability of a drug affects the conditions under which it can be distributed to pharmacists and stored by patients (e.g., Is it heat sensitive? Must it be refrigerated?). The metabolic stability determines how long the product remains effective in a patient's body and how much product must be administered to achieve a therapeutic effect. Metabolic stability also influences how many daily doses are necessary, and it can also be a factor in side effects. A drug's solubility affects numerous outcomes: if it will be stored in fat, whether or not it can travel through the bloodstream, if it can cross the gastrointestinal lining or be delivered through the skin, and if the kidneys can excrete it. These factors impact drug delivery, dosage regimens, and the potential for side effects. Toxicity or side effect limits may be relaxed if the alternatives to drug treatment warrant it.

Copying drugs is a sufficiently lucrative business to motivate companies to specialize in challenging patents and developing generic versions of drugs with expired patents (see *Generic Drugs* in Chapter 8). To prevent competitors from capitalizing on the efforts of pioneers and selling drugs for prices which reflect their reduced R&D burden, drugs must have some form of commercial protection. Drugs may be protected by patents, trade secrets, or other methods such as FDA-granted temporary market exclusivity. The vehicles that offer exclusive market rights for drugs—described in Chapters 7 and 8—are designed to promote innovation by granting drug developers temporary monopolies that permit them to recoup their investments in research and development.

Chapter 5

Tools and Techniques

> I think the evidence is overwhelming that one tool
> doesn't do it. Yet time and time again, we see new
> entrants coming into this business saying, "This tool will
> revolutionize the discovery process," when it's much
> more likely that the integration of tools, of how they
> work, will have a much more powerful effect.
> *Harvard Business School Professor Gary P. Pisano*

It is only through an understanding of the tools and techniques used for biotechnology research that one can develop an appreciation of the possibilities and challenges of biotechnology. This chapter presents a survey of biotechnology tools and techniques to foster such an appreciation.

The tools and techniques used for biotechnology research define the universe of products and services that biotechnology companies can develop. While a biotechnology company could decide to focus on a set of applications or a set of technologies, most define themselves around technologies rather than applications. From a practical perspective, it is simpler to focus on a technique that can have several applications than to search for all the methods to solve a single problem. Whereas tackling a specific application may require expertise in numerous techniques, a single technique can be used for multiple applications, positioning research and development to serve multiple markets reduces market risks associated with any individual market. Therefore, it is simpler and often preferable to develop expertise and patents for a few techniques and exploit this competitive advantage to develop solutions for multiple applications.

BIOINFORMATICS

Bioinformatics is the convergence of information technology and biotechnology, applying information technology to manage and analyze the vast amounts of data generated from basic biological research. Bioinformatics assists scientists in managing data and enables interpretation of data by presenting it in useful formats.

The tools and techniques that define bioinformatics are themselves a demonstration of the growth and diversity of techniques in drug discovery. As late as the mid-20th century, drug discovery was conducted mainly through chemical synthesis followed by extensive trial-and-error testing. Large-scale testing of derivatives of potential drugs was introduced in the 1970s, followed by attempts at rational drug design in the 1980s. Bioinformatics entered the arena in the 1990s, enabling drug synthesis and testing to be simulated by computers. As biological knowledge, computational power, and computer algorithms improve, it becomes increasingly possible to identify and refine potential leads through the use of computers.

There are two successive elements in bioinformatics: data assembly and data analysis. Computer-assisted data management enables gathering, analysis, and representation of biological information, helping scientists better understand biological processes, understand the mechanisms behind diseases, develop methods to treat diseases, and develop other applications based on biological knowledge. Bioinformatics also allows researchers to perform comparative and predictive studies of biological processes. Applications of bioinformatics data analysis include prediction of protein structure, prediction of protein function, and drug target selection.

One important development that emerged at the beginning of the twenty-first century was the implementation of automated research techniques such as DNA sequencing and robotic fluid-handling and assay systems, permitting large-scale research efforts. A defining event in the automation of biological research was sequencing the human genome. This mammoth project, extending over a decade, determined the sequence of the three billion base-pairs that comprise human DNA. The logistical problems of collecting and managing this mass of information required the development and application of novel computer technologies to assist biological research.

Sequencing the three billion base-pairs of the human genome was only possible with the use of bioinformatics to manage all the data. The information that can be processed using bioinformatics techniques includes not only sequence information for genes and proteins, but also details on the structure and function of proteins, disease correlations, and raw information produced from scientific experiments such as microarray analyses and protein interaction studies.

Bioinformatics applications exist for most steps of drug development. Predictive and analytical algorithms can screen potential lead compounds or help design them from scratch; toxicity can be predicted by comparison against compounds with known properties; even clinical trials can be simulated. Any biological information that can be entered into a computer database is subject to processing and analysis by bioinformatics.

A strength of bioinformatics is the ability to extract information and identify patterns from large databases. Data mining, an analytical bioinformatics application, uses computers to analyze masses of information. Integration and comparison of numerous experimental observations permits the discovery of trends and patterns in large databases, potentially identifying novel relationships.

CLINICAL MODELING

Clinical trials are necessary to demonstrate that drugs are safe and effective. Drug developers are not permitted to market drugs until they receive FDA approval, based on safety and efficacy data generated in clinical trials. Time spent in clinical trials is time that cannot be spent selling drugs. Because drug patents must be filed prior to the initiation of clinical trials, any reduction in clinical trial duration can be valuable—a single day's delay in approving a billion dollar blockbuster drug can mean a loss of revenue exceeding $3 million.

The processes for clinical trials are flexible—they can be adapted for diverse drugs—positioning clinical research organizations to build businesses around this critical step in drug development. By focusing on clinical trials, contract research organizations are able to develop specialized expertise and relationships with clinical trial providers. Clinical research organizations generally offer two distinct solutions to aid clinical trials. Trial management solutions involve strategies to design trials, facilitate patient recruitment, speed and improve communication between trial investigators and sponsors, and manage and analyze data. Simulation and prediction services use sophisticated computer techniques to enable the safety and efficacy of drugs to be predicted at a lower cost and with greater speed than actual clinical trials, ultimately permitting selection of compounds and trial protocols most likely to lead to FDA approval.

COMBINATORIAL CHEMISTRY

Combinatorial chemistry is a general method for creating a large number of molecules and systematically testing them for desired properties. An example of the automation of research, combinatorial chemistry is roughly analogous to a structured implementation of setting one million monkeys typing randomly on typewriters in the hope that at least one will produce a complete novel.

Combinatorial chemistry systems are designed to produce a variety of molecules according to a set of predefined rules. The library of synthesized molecules is then tested via high-throughput screening in a model system, possibly an enzyme-activity or protein-binding assay, to determine which compounds show promise for a given application. A combinatorial chemistry experiment may demonstrate interaction between two compounds or the ability of test compounds to enable or inhibit a chemical or biological reaction. Promising compounds may be used for further rounds of combinatorial chemistry, to search for related compounds with improved properties, or may be used directly as target compounds for further development and testing.

Variations on combinatorial chemistry such as combinatorial genetics apply the same general paradigm to different kinds of problems. Combinatorial genetics, a form of molecular evolution, involves the mutation of a gene to produce a library of variants. These variants are then analyzed for desired qualities. This technique has particular appeal for industrial biotechnology, where it can aid in the search for improved enzymes with improved yields, stability, or other characteristics.

FUNCTIONAL GENOMICS

Functional genomics seeks to understand the activities of genes in healthy and diseased states. The reality of human genetic variation means that different patients respond differently to the same drug. The difference may be as simple as a slight difference in efficacy or it may result in a drug being completely ineffective or even toxic in some patients. It is estimated that most commonly used drugs are effective in only 30–60 percent of patients with a given disease. A subset of these patients may suffer severe side effects.

Without functional genomics there is no simple way to determine if a given patient or subset of the population is likely to respond either well or poorly to a medication. As a result, drugs are developed for the "average patient." Furthermore, many drugs that might benefit a subset of patients may never be developed because they cannot be shown to be useful in an average group of patients. Patients who are unlikely to benefit from a drug or who may suffer adverse side effects are likewise not identified and given more appropriate treatments. Functional genomics enables segmentation of patient groups to resolve these issues.

Knowledge of the sequence of the human genome is a valuable tool for functional genomics. However, simply knowing the sequences of genes is not sufficient. Discrete genetic differences between individuals must be correlated with the effects of medications. Studying single nucleotide polymorphisms (SNPs) and pharmacogenetics reveals correlations that enable functional genomics.

SNPs are discrete DNA sequence changes between individuals that are at the root of many genetic differences. SNPs have been linked to the likelihood that an individual will find a drug effective or unsafe. These therapeutic differences are related to variations in drug targets, in enzymes that metabolize drugs, and in other molecules involved in cellular metabolism. The elucidation of discrete genetic differences that can be readily identified holds the potential to predetermine how a patient will respond to a drug.

PHARMACOGENETICS & PHARMACOGENOMICS

Pharmacogenetics studies the relation between genetic variation and the effects of pharmaceuticals: the investigation of how genetic differences affect the ways in which people respond to drugs. Specifically, pharmacogenetics seeks to understand the differences between drug targets and metabolic enzymes that affect efficacy and toxicity. Differences in genetic sequences are responsible for many of the differences between individuals. Just as genes influence eye color and hair color, they can also influence susceptibility to disease and determine whether specific drugs are safe and effective for certain individuals. The terms pharmacogenetics and pharmacogenomics are sometimes used to respectively distinguish between the correlation of single drugs with multiple genomes, and of multiple drugs with single genomes.

Learning why certain individuals are unresponsive to drugs or experience dangerous side effects gives researchers the potential to develop drugs that address these shortcomings. Studying the mechanisms by which drugs are rendered ineffective or toxic may also enable drugs to be designed to avoid or compensate for these alterations. Furthermore, drug discovery cost and time can be reduced by eliminating potential clinical trial participants for whom drugs in development are likely to

---

Box

Cytochrome p450 and pharmacogenomics

Cytochrome p450 is a generic term for a set of enzymes which are collectively the most important element in chemical modification and degradation of chemicals including drugs and other foreign compounds. A vast majority of the most serious adverse reactions to medicines appear to involve drugs that are metabolized by the cytochrome p450 system.[1]

Six different p450 genes are responsible for most of the metabolism of commonly used drugs. Each gene can have dozens of discrete mutations affecting its activity. Inventorying the set of cytochrome p450 enzymes and elucidating the factors contributing to their expression and activity levels is central to understanding and predicting differences in response to drugs. Some of the drugs and compounds degraded by cytochrome p450 enzymes are caffeine, morphine, Taxol, Prilosec, cocaine, codeine, Viagra, St. John's wort, and HIV protease inhibitors. If two compounds are degraded by the same cytochrome p450 enzyme it is possible that taking both compounds at the same time can lead one or both to accumulate to dangerous levels. This is one of the ways in which drugs can interact to alter efficacy or have lethal consequences.

Roche's AmpliChip 450 is the first microarray-based diagnostic test that can detect genetic variations influencing drug efficacy and adverse drug reactions. The AmpliChip contains 15,000 DNA sequences representing 31 genetic variations in two cytochrome p450 enzymes. According to Roche, the two enzymes affect 25 percent of commonly prescribed medications. The purpose of the chip is to determine whether a patient metabolizes drugs at a normal, slow, or fast rate. This information can help doctors prescribe appropriate medications and dosages based on a patient's rate of degrading specific drugs. Properly calibrated dosages can mean the difference between no response, therapeutic effectiveness, and serious side effects.

---

1 For a topical review, see: David A. Katz, D.A., Murray, B., Bhathena, A., Sahelijo, L. Defining drug disposition determinants: a pharmacogenetic-pharmacokinetic strategy. *Nature Reviews Drug Discovery*, 2008. 7:293-305.

prove ineffectual. More precise clinical trials justify smaller and fewer trials, facilitating FDA approval (see commentary on Herceptin clinical trials in Box *Personalized medicine and drug sales* in Chapter 6).

Functional genomics can facilitate drug discovery and improve drug administration. Applying functional genomics in a personalized approach to medicine—prescribing drugs only to patients likely to benefit from them—can streamline medical care and avoid unnecessary side effects. Functional genomics can also potentially identify patient groups of less than 200,000 Americans, qualifying drugs for Orphan Drug status (see *Orphan Drugs* in Chapter 8).

Specific challenges to implementing functional genomics are the development of tools to profile individual patients and retooling drug development for smaller patient groups. In order to prescribe drugs appropriately to an individual's genetic profile, a system to rapidly and inexpensively determine appropriate elements of an individual patient's genetic profile is necessary. The gains in safety and efficacy of drugs with functional genomics-based prescription are coupled with smaller patient populations. Accordingly, drug firms must be willing to actively exclude patients in order to realize the benefits of functional genomics. Smaller patient populations don't necessarily mean smaller profits—see the Box *Genzyme: Building an empire on orphans* in Chapter 8.

MICROARRAYS

Microarrays are tools that enable the identification of DNA or other samples and examination of gene expression and protein modifications in individual tissues, and under different conditions. While the first microarrays were directed at detecting DNA, new technologies have enabled protein-based and other forms of microarrays. Microarrays enable researchers to detect the presence or expression of many genes and proteins at once. The ability to simultaneously examine the changes in expression of many different genes, or changes in protein levels and modifications, is useful in investigating the effects of diseases, environmental factors, drugs, and other treatments in human health.

Applications of microarrays include diagnosing or identifying cancerous cells, assessing genetic predispositions to diseases, examination of gene expression, and gene and protein responses to drugs or other therapeutic procedures.

PROTEOMICS

Proteomics is the study of protein structure and function. Genes encode proteins, which perform structural and functional roles in cells. It is proteins, not genes, which are the major actors in molecular biology (see Chapter 3). Understanding the structure and function of proteins can lead to new therapies and influence disease diagnosis and treatment.

Proteins can be roughly categorized by their structural and functional roles. An example of a structural protein is keratin, a component of skin. Functional proteins, called enzymes, perform cellular duties. Metabolic enzymes aid in food digestion and enable harvesting of stored energy. Studying DNA can reveal some information on the control of protein synthesis, but provides limited information about the structure and function of proteins. This requires examination of the proteins themselves.

Proteomics uses a variety of techniques to examine protein structures and functions. Unlike DNA sequencing, which is a relatively uniform technique that can be widely used without modification, the very methods used to investigate proteins vary with each individual protein being studied. There is no simple or uniform way to produce, identify, quantify, or characterize proteins. The need to continually adjust experimental methods in proteomics research is a significant challenge to scaling or automating research efforts.

MANUFACTURING

Developing a drug that is safe and effective is essential to gain FDA approval, but to generate revenues and recoup development costs it is necessary to manufacture and sell the product as well. Conventional wisdom once held that investing in manufacturing process development did not benefit a company's returns as much as basic research. In an era of increased competition where companies frequently produce competing treatments, the ability to accurately predict demand, to rapidly develop production methods, and to scale production capacity provide a strategic advantage.

In the process of drug development a drug may be produced in test tubes, flasks, small fermentation vessels, pilot-plants, and large-scale production facilities. The progression from benchtop production to large-scale production is not a trivial process. As the scale of production increases, factors such as temperature, oxidation, and mixing change, potentially altering the final product. To ensure drug quality, manufacturers must demonstrate compliance with FDA current good manufacturing practices (cGMP) and further prove that drugs of consistent purity and activity can be produced in large quantities from batch to batch, day after day, year after year.

One alternative to traditional manufacturing methods is the use of animals and plants that are genetically modified to produce a desired compound. For example, drugs may be harvested from chicken eggs or cow milk, or purified from plant tissues. A significant benefit of transgenic production is that production of raw material is relatively simple to implement and scale. It is estimated that plant-based biologic production can be 10 to 1000 times less expensive than conventional fermentation systems. Leveraging the relative simplicity and cost advantage, drug-producing varieties of animals and plants can be distributed in regions lacking sufficient expertise, facilities, or resources for conventional fermentation production. To scale production, these transgenic factories can be bred or cloned, increasing the supply of raw product.

DRUG DELIVERY

Despite the emphasis on the biological activity of drugs, it is important to also consider the systems and methods used to deliver drugs to their therapeutic targets. The goal of drug delivery systems is to enable active medications to reach appropriate parts of the body, in the appropriate concentrations for the appropriate amount of time, where they can accomplish their therapeutic task.

Drugs produced by biotechnology techniques tend to be large proteins and nucleic acids which face special challenges relative to smaller, more chemically stable pharmaceutical drugs. Factors impeding oral delivery of biologic drugs include:

- Acidity of the digestive system
- Intestinal enzymes that degrade proteins
- Inability of biologic drugs to cross intestinal walls
- Poor solubility of biologic drugs

Overcoming the challenges of biologic delivery can present new opportunities, as targeted and metered dosage systems can potentially improve drug effectiveness, mitigate safety and side effect concerns, and ultimately improve patient compliance and retention.

Patient compliance is also a concern in drug delivery. The requirement to take many pills a day, to follow rigorous dosage regimens, or the use of unappealing delivery methods such as injection may discourage compliance, and patients cannot benefit from drugs they don't take. Reducing administration from several times a day to once a week by using an extended-release formulation, for example, can dramatically improve compliance and ultimately improve patient outcomes—the primary objective of drug therapy.

Delivery techniques that increase compliance can ultimately help a company derive more revenue from a product. Selling twice as much of a drug by doubling the duration that patients take the drug or doubling the number of people who take it is arguably similar to selling two drugs, without the cost of developing two drugs.

NANOTECHNOLOGY

Nanotechnology is a multidisciplinary field encompassing the development and application of materials at sizes measured in billionths of a meter. Surface tension, molecular interactions, and surface area exposure play an increasingly important role in chemical and physical interactions at this size range, giving nano-scale materials properties that are markedly different than those seen at larger scales. Nano-sized flour particles, for example, are capable of igniting violently and causing explosions in flour mills. Geckos are able to climb walls because of nano-scale hairs on their feet which use atomic interactions, rather than stickiness, to adhere. The enzymes and other key molecular players in biotechnology also operate at this size scale, creating an opportunity for convergence between biotechnology and nanotechnology.

As with biotechnology, early investments in nanotechnology research were attracted by the high profit potentials of serving unmet medical needs. Beyond developing new therapeutic products with nanotechnology, there are also strong opportunities in developing delivery systems that can improve the safety and efficacy of existing drugs (see *Drug Delivery* earlier in this chapter). Reducing the particle size of drugs has the potential to:

- Increase surface area
- Enhance solubility
- Improve oral bioavailability
- Speed onset of therapeutic effect
- Decrease necessary dosage
- Decrease variability between fed and fasted dosage
- Decrease patient-to-patient variability

Table 5.1 *Selected nanotechnology applications in drug delivery*

Technology	Benefit
Carriers	Improve solubility and avoid need for harsh solvents
Encapsulation	Extend duration of drug bioavailability
Nanoparticles	Improve solubility, speed delivery, and extend duration of drug bioavailability

Elan Pharmaceuticals' NanoCrystal technology overcomes solubility problems by using a proprietary technique to absorb nano-scaled particles of drug substance onto the surface of stabilizers. NanoCrystal technology allows drugs to be released into the bloodstream faster than alternative methods and enables drugs to stay dissolved for longer periods of time. In August 2000 Wyeth's solid-dose form of the immunosuppressant Rapamune became the first FDA-approved drug incorporating NanoCrystal technology. Rapamune was previously available only as an oral solution, requiring refrigerated storage and mixing with water or orange juice prior to administration. NanoCrystal tablets enable a solid dose formulation which does not require refrigeration.

Abraxane is an example of how nanotechnology can improve existing drugs. It was developed as an improved version of Taxol. The castor oil carrier required to solubilize Taxol (see *Drug Delivery* earlier in this chapter) has several significant side effects associated with it. By binding paclitaxel (a generic version of Taxol) to nano-scale protein particles, Abraxis Oncology produced a version of the drug that could be injected without castor oil, dramatically improving the side effect profile.

Nanotechnology also has many applications in research. Researchers continually seek to perform assays using smaller and smaller quantities of experimental materials, which are often very expensive and difficult to obtain. Smaller-scale experiments provide two advantages: the ability to perform more experiments at lower costs, and the ability to run more experiments in less space. Microarrays, for example, enable researchers to examine many genes at once using a small quantity of material. Nanotechnology-based solutions, such as "lab on a chip" products, can reduce the scale of experiments even further, enable multiple sequential or simultaneous assays, and expand opportunities for automation in research.

Chapter 6

Applications

The best way to predict the future is to invent it.
Richard Feynman

Biotechnology companies focus on selling products or offering services. Products include drugs, reagents, research tools, industrial enzymes, and specialized crop plants. Services include discovering drug lead compounds, clinical trial management, and manufacturing. More details on biotechnology business models are provided in Chapter 10.

The most common application for biotechnology companies is drug development. This is partially due to the enormous profit potential of drugs, which can greatly offset the increased development cost relative to other applications. Some firms specialize in elements of the drug development process, supporting the search for potential drug compounds by coordinating clinical trials or producing research tools and drug delivery systems. These "pick and shovel" companies benefit indirectly from the profits of other development companies by selling necessary products and services. While some service firms fulfill functions that could be developed internally, dedicated service firms may also possess economies of scale, enabling them to offer specialized expertise and abilities.

Drug development differs from most other commercial development ventures because products must be proven safe and effective before they can be marketed. Applications not intended for human use benefit from not requiring clinical trials to prove

Table 6.1 *Biotechnology application categories*

Category	Description
Green: Agricultural biotechnology	Products and applications related to livestock and crop production, and agricultural production of biotechnology products.
White: Industrial biotechnology	Modification or improvement of industrial processes such as paper processing, bioremediation, and chemical and organic compound synthesis.
Red: Medical biotechnology	Drugs and other agents to treat, cure, or prevent disease, and products that assist in the diagnosis of diseases or measurement of crucial factors in health and disease.

their safety and efficacy prior to commercial release, although these applications may still be subject to EPA, USDA, and other political or ethical restrictions. The regulatory controls influencing biotechnology are discussed in detail in Chapter 8.

Table 6.1 shows the division of biotechnology applications into three categories: Green biotechnology for agricultural applications, white biotechnology for industrial applications, and red biotechnology for therapeutic applications. Specific applications are described in further detail in this chapter.

In reading this chapter, it is important to recognize that biotechnology is not a panacea. In addition to practical considerations such as technological constraints, the legal, regulatory, political, and commercial factors described later in this book have profound impacts on the ability to develop and commercialize biotechnology. Many biotechnology companies fail because they develop products for which profit-enabling markets do not exist.

GREEN BIOTECHNOLOGY: AGRICULTURE

The directed modification of plants and animals can increase their value in agricultural applications. By studying the genes responsible for specific traits, it becomes possible to introduce, alter, or change the expression of those genes in a controlled manner, resulting in a desired change. Extensive testing, mandated by the FDA, EPA, and USDA, is required to determine if genetically modified plants are safe for humans and ensure that they do not pose a threat to the environment. This testing requires demonstration that foreign proteins in edible crops are decomposed by cooking or stomach acids, precluding their ability to cause adverse effects if ingested.

Traditional agriculture relies on crossbreeding and hybridization to improve the quality and yield of crops and domesticated animals, and to overcome natural obstacles such as disease. These methods involve controlled breeding of plants and animals with desirable traits to produce offspring that ideally will retain the best traits of the parent organisms. Virtually every plant and animal grown commercially for food or other uses is a product of crossbreeding and/or hybridization. Relative to biotechnology methods, these processes are costly, time consuming, inefficient, and subject to significant practical limitations.

Genetic modification of crops has produced herbicide resistant strains, insect resistant strains, enriched foods, and improved industrial products. Herbicide and pest resistant crops can have a profound positive impact on the environment by making it possible to raise crops using dramatically less pesticide. Transgenic corn containing insecticidal toxins from *Bacillus thuringiensis* (Bt) bacteria can prevent corn borer infestations without chemical crop dusting that is toxic to humans and also kills beneficial insects. Producing these herbicide and pest resistant crops by traditional methods—if it were possible at all—would take dozens of generations.

A study of the first ten years of commercial genetically modified crop growth—from 1996 to 2006—found economic benefits to farms of $5 billion in 2005 and $27 billion over the ten year period. Pesticide use was also reduced by 224 million kilograms, or 6.9 percent, resulting in a reduction of environmental impact by more than 15 percent.[1] While improved seeds

---

1 Brookes, G. and Barfoot, P. Global Impact of Biotech Crops: Socio-Economic and Environmental Effects in the First Ten Years of Commercial Use. *AgBioForum*, 2006. 9(3):139-151.

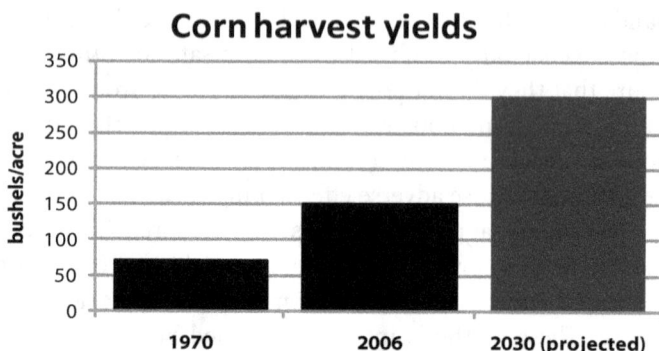

Figure 6.1 *Progress in agricultural yields*
Source: Monsanto

may cost significantly more than conventional seeds, it is estimated that conventional farmers spend significantly more on chemical insecticides and herbicide than they spend on seeds. Furthermore, genetically modified seeds have helped improve yields significantly; U.S. corn harvest yields have doubled since 1970, and Monsanto predicts that harvest yields will double again by 2030 (see Figure 6.1).[2]

The most abundant genetically modified crops are cotton, corn, soy, and canola. More than 1 billion acres of genetically modified crops had been sown in 17 countries by the end of 2005, a decade after Monsanto introduced the first genetically modified crop. According to the USDA, 52 percent of all corn, 79 percent of upland cotton, and 87 percent of soybeans planted in the United States in 2004-05 were biotechnology-derived varieties. The International Service for the Acquisition of Agri-Biotech Applications reports that in 2007 the number of farmers growing biotechnology-derived crops exceeded 12 million—11 million of whom were defined as resource-poor farmers—and hectarage exceeded 114 million acres.

---

2 Hindo, B. Monsanto: Winning the Ground War. *BusinessWeek*, December 17, 2007.

CHALLENGES

Techniques to insert genes into plants are well established, but a remaining challenge for agricultural biotechnology is the realization of desired modifications. With many plants sporting genomes larger than humans, a fundamental understanding of agricultural biology is necessary for application. Confounding efforts at genetic modification, introduced genes may be unstable, and unforeseen biological issues may interfere with introduced proteins.

Another challenge for agricultural biotechnology is public acceptance. Because public support is essential to enable application of biotechnology, it is important to be sensitive to potential objections and to encourage positive perceptions. Public concerns can impact political decisions, resulting in bans on crop plantings or even sales of crops. Figure 12.1 in Chapter 12 shows the dramatic effect of an EU-wide ban on genetically modified crops on U.S. corn exports.

Critics of genetically modified foods warn that inserting genes into plants and animals may have unforeseen results, increasing the risks of allergic reactions to foods and other health problems. The potential for genetically modified plants to crossbreed with wild stocks or to cause environmental damage is also a prevalent concern. It is worth noting that many of these same concerns would also preclude the use of traditional farming practices and breeding methods if they were applied beyond biotechnology.

Beyond answering critics of genetic modification, developers must also find markets with a preference for their products and who are also willing and able to pay a profitable price. A case example of failure to find a profit-enabling market is the Flavr Savr tomato, a tomato engineered to have a longer shelf-life. In 1994 the FDA approved the Flavr Savr tomato, the first genetically modified whole food product. Interestingly Flavr Savr tomatoes were pulled from the market not because of consumer resistance, but rather due to customer disinterest and an inability to sell for a profit.

While improved yields, nutritional enhancement, and the ability to grow crops on marginal soils may appear to individuals in developed nations to be trivial or cosmetic improvements, but the situation in developing countries makes these improvements far more imperative. As population encroachment and environmental change are decreasing the quantities of arable land in many parts of the world, aging distribution infrastructures are challenged to deliver food to ever-increasing populations. For many developing nations the ability of biotechnology to increase crop yield, reduce farming inputs, and expand arable land presents a vital solution to the need to grow more crops on less land in order to prevent otherwise-inevitable widespread starvation.

TREE BIOTECHNOLOGY

An early goal of forest scientists has been to produce trees with less lignin, avoiding expensive and environmentally toxic procedures in paper production. Removal of lignin in paper production requires an enormous amount of energy and chemicals, making the pulp and paper industry the second most energy-intensive industry group in the U.S. manufacturing sector. Another problem is that forestry demand for wood products exceeds the rate of renewal. Some tree species require well over a century to reach economic viability. Using genetic engineering to develop faster, straighter, and taller-growing trees can potentially fill a market need while preserving old growth forests.

ANIMAL BREEDING AND CLONING

Biotechnology can serve two roles in animal breeding. First, it can enable the use of genetic markers to identify desired animals for breeding programs. Biomarkers associated with characteristics such as milk production, meat quality, or hereditary diseases can be used to inform traditional breeding programs and improve herd quality. The second, more contentious application, is to directly clone desired animals.

In 2008 the FDA released a "final risk assessment," concluding that foods from healthy cloned animals are as safe as those from traditionally-bred animals. Cloning animals costs substantially more than traditional or assisted breeding programs, but this additional cost is somewhat offset by the near-certainty that the clone will inherit desired traits. The extra cost of producing cloned animals also provides an assurance to wary consumers—the high cost of producing these animals means that they are unlikely to be used as meat or milk sources, but will likely be relegated to breeding.

FUNCTIONAL FOODS

Functional foods contain elements which can provide health benefits beyond simple nutrition. Existing examples include high omega-3 eggs, produced by feeding ground flax seeds to hens, and supplementing processed foods with soy protein or fiber.

Beyond improving yields, biotechnology also has the potential to dramatically impact the nutritional qualities of food. Nutritional modifications to plants include conferring the ability to synthesize essential vitamins, reducing the undesirable saturated fat content of cooking oils, increasing protein quantity and quality in vegetable staples, and reducing allergenic properties of milk, wheat, and other products.

These nutritional improvements can have a significant effect on human health. For example, enabling staple crops such as rice to synthesize vitamin A precursors or to make iron bio-available respectively hold the potential to prevent blindness and anemia in countries where commercial vitamin supplements are unaffordable. Calgene's (now owned by Monsanto) Laurical is the first commercially available functional food oil, approved by the FDA in 1995. While conventional canola oil does not contain lauric acid, Laurical contains 38 percent lauric acid. Laurical has applications in soaps and detergents, chocolate, low-fat coffee whiteners, and imitation cheeses.

MOLECULAR FARMING

Molecular farming produces useful products from domesticated plants and animals through genetic engineering.

Box

Carnivorous fish as vegetarians

Fish farming, or aquaculture, has been heralded as a potential solution to overfishing of wild stocks. The situation is complicated, however, by the need to satisfy carnivorous fish diets. Carnivorous fish aquaculture accounts for the majority of global fish oil usage, and is rapidly growing to become the primary market for fish meal as well—aquaculture is depleting wild fish stocks.

Enter biotechnology. By studying the protein requirements of carnivorous fish, researchers can potentially use traditional breeding and genetic modification to produce terrestrial crops as feed. These crops can both address the nutritional needs of fish, and can resolve some of the environmental issues related to aquaculture. Furthermore, growing plant crops to feed fish farms creates new revenue opportunities for farmers and holds the potential to enable inland farmers to farm fish, improving the distribution and availability of fresh fish.

Despite its promise, it remains to be seen whether traditional breeding can produce plants that can satisfy the nutritious needs of fish, if consumers will accept fish products that have been genetically modified plants, if the quality and flavor of plant-fed fish will match fish-fed or wild fish, and if plant-fed aquaculture can be economically feasible.

Pharming is a subset of molecular farming and produces therapeutic drugs using genetically altered animals and plants. The distinction between molecular farming and traditional farming is that the plant and animal products of molecular farming are not eaten as food, but are harvested to produce useful biotechnology products. Safeguards to prevent exposure through accidental ingestion include sequestration of crops (plans have included growing plants in abandoned mine shafts), production of recombinant proteins in non-edible portions of plants, and expressing proteins at very low levels, requiring extensive processing to obtain measurable and useful quantities of recombinant materials.

A non-therapeutic application of molecular farming is the mass-production of spider silk. Stronger than steel and lighter than cotton, spider silk manufacturing has traditionally been impeded by the inability to domesticate spiders. Nexia Biotechnologies has produced dragline spider silk in laboratory conditions. Their intention is to spin spider silk produced in the mammary glands of genetically modified goats for use in fishing line and in military applications such as lightweight body armor.

Pharming

Many developing countries that could benefit from commercially available therapies for diseases such as hepatitis and other endemic conditions cannot afford to purchase appropriate medicines or produce them locally. Production of therapeutic vaccines in familiar crops such as bananas and potatoes, or chicken eggs, can enable local farmers to manufacture medicines without the need for sophisticated production techniques or expensive purification methods.

The traditional method for manufacturing biological drugs is fermentation in huge stainless steel vats. A significant advantage of pharming is that it can decrease the cost of drug manufacturing. Furthermore, whereas scaling fermentation systems requires building and receiving FDA approval for additional facilities, boosting pharmed drug production may be as simple as planting more transgenic plants or increasing the size of a transgenic animal herd. This is especially important in countries where expertise and facilities for large-scale fermentation are not available. These advantages are offset by the higher up-front costs and longer lead-times required to produce transgenic animal or plant production systems (see *Manufacturing* in Chapter 5).

The development of drugs that are easy to purify or that can be administered without purification is essential for enabling pharming. Application of pharming is challenged by the threat of uncontrolled spread of genetically engineered plants and animals, and by start-up development costs. In the case of drug manufacturing, it is preferable to retain formulation flexibility and delay the bulk of manufacturing expenditures until a drug's safety and efficacy have been assessed. The financial cost and time required to develop transgenic production systems are at odds with this strategy.

WHITE BIOTECHNOLOGY: INDUSTRIAL PROCESSES & BIO-BASED PRODUCTS

Industrial biotechnology is the application of molecular biology techniques to improve efficiency and reduce the environmental impacts of industrial processes. Just as biotechnology has transformed agriculture, drug discovery, and development, it can similarly affect industrial operations.

Industrial biotechnology companies develop biocatalysts such as enzymes that are used for chemical synthesis. Enzymes are a category of proteins which are produced by all living organisms (see *Proteins and Enzymes* in Chapter 3). Enzymes enable the biochemical reactions necessary for life by increasing reaction rates. In biological systems, enzymes help digest food, assemble complex molecules, and perform other complex functions. Specialized enzymes are also used extensively as detergents as well as in the production of beer, cheese, and fruit juice. Bacteria have developed specialized enzymes that allow them to live in a wide variety of extreme environments; from thermal vents at the bottoms of oceans to the insides of rocks. Enzymes are characterized according to the compounds they act upon. Some of the most common enzymes with industrial applications are proteases, which break down protein; cellulases, which break down cellulose; lipases, which act on fatty acids and oils; and amylases, which break starch down into simple sugars.

By studying diverse bacteria and other organisms, scientists discover novel biocatalysts that function optimally under a wide variety of conditions, including the relatively extreme levels of acidity, salinity, temperature, or pressure found in some industrial manufacturing processes. In other cases, enzymes can remove the need for extreme conditions or harsh chemicals, saving energy and reducing environmental impact.

The application of biotechnology to industrial processes is appealing because of the potential to affect yield, effective-

> **Box**
>
> ## Blue jeans and biotechnology
>
> 1.8 billion pairs of denim jeans are sold each year, making them among the most prevalent clothing items sold worldwide. *Stonewashing* is commonly used to soften the jeans and fade the dyes to give the jeans a slightly worn appearance. This process was traditionally performed by tumbling jeans in large machines with abrasive pumice stones. This process can weaken jeans and damage washing machines, and requires several rinsings to remove all the pumice traces.
>
> An alternative enzyme-based method has been introduced which imparts several benefits. The degree of stonewashing can be attenuated by using cellulase enzymes to break down the denim cellulose fibers in a controlled manner. This process also requires less water and energy than traditional stonewashing, and results in longer-lasting jeans.

ness, and production cost of products with established markets. Serving established markets cam improve the accuracy of market size projections, helping justify high R&D investments and attracting funding for large market opportunities. The potential for application of biotechnology to reduce infrastructure requirements may also make it possible to profitably address smaller markets. For example, see the example of *Oil Well Completion* below. An additional appeal for industrial biotechnology development is the greatly reduced regulatory burden relative to pharmaceutical applications.

BIOFUEL AND LUBRICANTS

Petroleum prices, political considerations, and the threat of shortages all motivate the search for alternative fuel sources. Processes to convert cornstarch into ethanol, a petroleum additive and alternative, have been available for many years, but questions about the scalability and ultimate economics of this approach have led to the search for alternative methods to produce ethanol.

Fuels derived from petroleum are the product of compression and heating of prehistoric plants and animals over geological time scales deep below the earth's surface. Because of their ultimate biological source, the potential exists to use alternative processes to make fuels from sources such as plant materials and animal fats in less time. There are three basic methods to produce fuels from plant and animal sources: chemical transesterification, fermentation, and cellulose degradation. These fuels can be used in cars and trucks, as well as in numerous other applications.

Transesterification is a process used to convert vegetable oils, animal fats, and recycled greases into biodiesel. This is technically a chemical process, not an application of biotechnology. Fermentation is the use of bacteria or yeast to convert simple sugars, abundant in plants such as corn and sugar cane, into ethanol. This is fundamentally the same process that is used to make beer, wine, and other alcohols. Cellulose degradation significantly expands the prospects for fermentation by enabling the use of a wide variety of feedstocks. Whereas fermentation requires feedstocks with a high content of simple sugars, cellulose degradation uses chemical pre-treatments and cellulase enzymes to break down cellulose, a complex sugar, into simple sugars. These simple sugars can then be used in traditional fermentation to make desired chemicals.

The principal advantage of cellulose degradation over other methods is the abundance of cellulose in materials such as farm waste, wood chips, and even garbage. A majority of the material in plants is cellulose. Cellulose degradation occurs in nature, but slowly. The challenge is to increase the efficiency of cellulase, an enzyme that breaks down cellulose, and to improve the yield of cellulose from biological sources.

Subsidies on corn production and tax exemptions for non-petroleum fuels have been instrumental in enabling entrants to produce and market biofuels. The situation for biofuels is analogous to the early years of penicillin production. The basic scientific principles are known, a strong market need exists, but a better method is needed to enable cost-effective large-scale production.

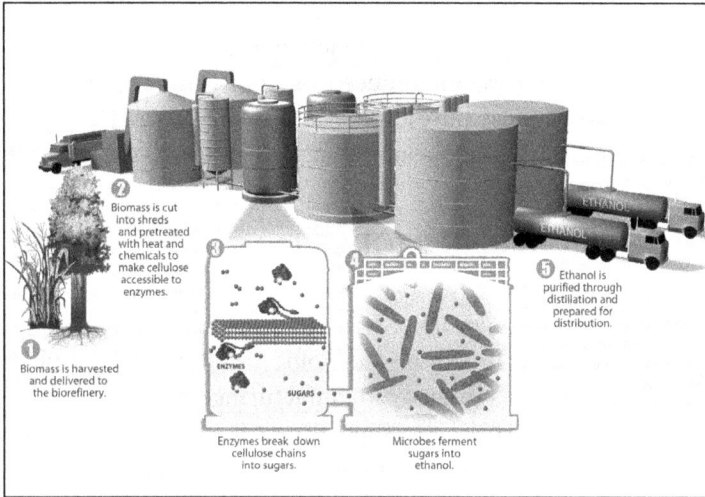

Figure 6.2 *How cellulosic ethanol is made*
Source: Genome Management Information System, Oak Ridge National
    Laboratory

PLASTIC

The world's first modern biorefinery, a Cargill-Dow project, went online in Blair, Nebraska in 2002. The plant is the product of a joint venture established in 1997 between Cargill and the Dow Chemical Company to commercialize polylactic acid (PLA) under the brand name NatureWorks. Dow pulled out of the venture in 2004 acknowledging significant long-term potential, but dissatisfaction with short-term profitability.

PLA is made by fermenting the sugar in corn (other high-sugar feedstocks are also amenable to this process) into lactic acid molecules, which are then linked to form polylactic acid. PLA can be used to make a wide array of products, including plastic cups and containers, wrappers, and polyester textiles. Furthermore, PLA is biodegradable, requires 65 percent less energy to produce than conventional plastics, and can reduce fossil fuel use in plastic manufacture by up to 80 percent.

Bio-based plastics have the added benefit of being naturally biodegradable, reducing the environmental impact of their use. They face resistance due to their higher costs, the need to re-engineer downstream manufacturing and utilization processes in some cases, and reduced suitability in harsh environments. Despite these hurdles, they are finding strong adoption in consumer-facing applications such as disposable packaging for food containers.

OTHER BIO-BASED PRODUCTS

As described in the section *Biofuels* above, the original source of petroleum products is actually biological. Accordingly, the potential exists for biotechnology innovations to replace petroleum products in many manufacturing processes.

Vitamin B2 is used as a supplement in animal feed to keep animals healthy and fit. In 1990 BASF developed an innovative fermentation method to replace the traditional eight-step chemical process used for vitamin B2 production, using *Ashbya gossypii* fungus with a one-step fermentation. This fermentation process reduces costs by up to 40 percent and reduces environmental impact by 40 percent. The fermentation process has several other advantages over chemical synthesis. BASF has realized a 95 percent reduction in waste, reduced energy usage due to lower reaction temperatures, and a 60 percent reduction in the resources required.

Another product improved by biotechnology is propanediol. Propanediol is a clear colorless liquid with applications in deicing, as an engine coolant, in adhesives and coatings, and as an additive in cosmetics and shampoo. Traditionally produced from petroleum feedstocks at high temperatures, a joint venture of Dupont and Tate & Lyle uses a proprietary fermentation method to produce a biologically-derived version of propanediol named Bio-PDO for industrial and consumer applications. The benefits of Bio-PDO over conventional alternatives are reduced production energy requirements, low toxicity, and biodegradability, and improved heat stability and reduced corrosion when used as anti-freeze. Bio-PDO is also replacing petroleum sources in the manufacture of Dupont's Sorono plastic.

DETERGENTS

Detergent enzymes represent the broadest application of enzymes. Detergent enzymes improve household laundry, dishwashing, and industrial washing applications by improving cleaning performance, reducing washing times, reducing energy consumption by lowering wash temperatures, and even rejuvenating clothes.

The most common enzymes used are proteases and amylases, which respectively remove stains and soils based on proteins and starches. Other enzymes with applications as detergent adjuncts include lipases to digest fat or oil based stains, peroxidases to inhibit dye transfer, and cellulases to prevent pilling on cotton clothes.

These innovations allow a reduction in the use of numerous environmentally-damaging chemicals, including phosphates and bleaches.

MINING

Microorganisms are used worldwide in mining processes to oxidize and leach metals. Other applications are as alternatives to harsh chemicals to remove metals from industrial wastewater streams. The primary bacteria employed are *Thiobacillus fer-*

Table 6.2 *Selected industrial enzymes*

Enzyme type	Function and utility
amylase	Decomposes simple sugars. Applications in textiles, laundry and dishwashing, biofuels, and paper production.
cellulase	Decomposes cellulose into simpler sugars. Applications in biofuel production, laundry, and paper processing.
lipase	Decomposes fats. Applications in laundry and surface cleaning, food processing, leather processing, and pharmaceuticals.
protease	Decomposes proteins. Applications in laundry, leather processing, baking.
xylanase	Degrades plant cell walls. Applications in paper production, biofuels, food production, and textiles.

rooxidans, Leptospirillum ferrooxidans, and thermophilic (high temperature) bacteria to leach metals such as copper and gold from sulfide minerals. Some of the advantages of bioleaching over conventional roasters, smelters, and pressure autoclaves are that construction time is shorter, no noxious gases or toxic effluents are produced, environmental permit and reporting processes are simpler, and safety is increased due to processing at or near ambient temperatures and pressures.

Another method under development is the use of plants to mine sparse deposits of valuable minerals. Plants with enhanced abilities to sequester heavy metals in soil can extract sparse deposits of gold or other valuable minerals, which can then be recovered by simply harvesting and incinerating the plants.

OIL WELL COMPLETION

Microbial enhanced oil recovery (MEOR) is the use of microorganisms to retrieve recalcitrant oil from existing wells, maximizing petroleum production of an oil reservoir. MEOR employs the inoculation of selected natural bacterial strains into oil wells to decrease the viscosity of thick oil deposits and ease oil flow, or to produce gases such as carbon dioxide to propel oil out of the well.

BIOREMEDIATION

Bioremediation is the application of biotechnology for environmental reclamation. Some of the processes described above have applications in bioremediation. Relative to existing alternatives, the use of plants, microorganisms, and their by-products to sequester pollutants, or to degrade them into relatively benign compounds, can be a safer, cheaper, and faster method to clean the environment.

Unresolved questions regarding the release of genetically modified organisms into the environment motivate the search for more natural techniques. Fortunately, there are few natural materials that at least one naturally-ocurring microorganism cannot use as a nutrient. Given appropriate conditions, even synthetic compounds are subject to microbial metabolism. By

Box

Using bacteria to make snow

Snowmax is an ice-nucleating protein derived from naturally-oc-curring *Pseudomonas syringae* bacteria. It is hypothesized that the natural purpose of this ice-nucleating protein is as part of a long-distance dispersion strategy of *Pseudomonas syringae*, which is a plant pathogen. The bacteria are able to survive for long periods in aerial suspension. The ice-nucleating proteins help drop the bacteria out of aerial suspension, by way of rain or snow, enabling them to infect plants.

Specially-designed aeration guns are used to spray water mixed with Snowmax on ski slopes. Snowmax increases the number of nucleation centers in water droplets from these aeration guns, improving snow making efficiency and also enabling snow making at higher temperatures, ultimately saving ski resorts money and improving the quality of ski slopes. Snowmax is produced by growing *Pseudomonas syringae* in a fermentation vessel and extracting the protein using filtration processes. The product is irradiated prior to shipping to ensure that live bacteria, which are not harmful to humans, are not released.

searching for organisms already feeding on pollutants, either in natural environments or at polluted sites, it is possible to develop non-transgenic bioremediation systems with applications ranging from treatment of oil spills to reclamation of contaminated soil and water.

RED BIOTECHNOLOGY: MEDICAL APPLICATIONS

MONOCLONAL ANTIBODIES

Antibodies are natural components of human and other immune systems that recognize unfamiliar material such as infectious bacteria and cancerous cells and help eliminate them. While our natural complement of antibodies is generally very effective at recognizing and prompting the destruction of infectious microorganisms and cancerous cells, threats are sometimes missed. Monoclonal antibodies are a category of biotechnology-derived drugs that are designed to act and look like naturally occurring antibodies and may directly treat diseases or condition a patient's own immune system to launch a highly specific attack on infections or diseased tissues. They are designated "monoclonal" because they are produced as large batches of identical molecules. Georges Köhler and César Milstein received the 1984 Nobel Prize in Physiology or Medicine for their description of a technique to produce monoclonal antibodies. They shared the prize with Niels Jerne, who described the development and control of the immune system.

In the 1980s the first antibody trials saw early experimental therapies rendered inactive by the liver, or activating patients' own immune systems to raise antibodies against the foreign therapeutic antibodies, resulting in increased illness. The rejection of these initial non-human antibodies can be attributed to the primary purpose of the immune system: to repel foreign bodies. The use of modified versions of animal antibodies, humanized antibodies, and fully human antibodies led to the development of monoclonal antibody therapies that are safe and effective.

Genentech's Rituxan was the first monoclonal antibody to be approved for cancer treatment in the United States. Rituxan works by binding to specific types of cancer cells and triggering the immune system to destroy them. Another Genentech product, Herceptin, is targeted at growth factors that are directly implicated in approximately 20 percent of breast cancers (see Box *Personalized medicine and drug sales* later in this chapter).

Antibody-based drugs can enjoy years of strong sales with minimal competition, even after patent expiration, because of the difficulty of demonstrating equivalence of antibodies produced by a second party. This challenge, however, motivates competitors to produce improved antibodies rather than simply producing competing antibodies. This means that when an innovator's antibody is challenged by a new entrant, the innovator is likely to lose a greater portion of market share than they might if the competing drug were merely equivalent to the original.

RNA INTERFERENCE

RNA interference therapies, sometimes referred to as anti-sense therapies, block gene expression. Summarizing information presented in Chapter 3, mRNA is a molecule that transfers information from genes to the protein synthesis machinery within cells. The goal of RNA interference is to intercept an mRNA message before it is translated into protein. Andrew Fire and Craig Mello shared the 2006 Nobel Prize in Physiology or Medicine for their discovery of gene silencing by RNA interference.

It is important to recognize that RNA interference is a subtractive solution. Interference cannot directly replace, amplify, or add a gene function. It can only inhibit a gene function (although inhibiting an inhibitory gene can indirectly increase the expression of a second gene). RNA interference has the potential to treat a range of diseases including cancer, autoimmune disorders, and infectious diseases.

A successful RNA interference therapy must enable entry of interfering molecules into cells to permit a therapeutic effect, prevent degradation of interfering molecules before they can act, and ensure specificity so that essential functions are not disrupted. The relative ease of controlling these issues in laboratory settings means that many RNA interference therapies that look promising in pre-clinical development are likely to face complications in therapeutic settings. Isis Pharmaceuticals' Vitravene, the first RNA interference drug, overcomes delivery issues through direct injection into the target tissue.

GENE THERAPY

Gene therapy uses genes to treat disease. Techniques for gene therapy include replacement of defective genes, and supplementation with therapeutic genes. For diseases caused by an absent or defective copy of a specific gene, supplementation of that gene can potentially cure the disease. The technical challenges of gene therapy include targeting appropriate cells and tissues, ensuring gene transfer, controlling gene expression, and satisfying safety concerns.

While most genetic deficits require gene expression in specific cell types, some diseases can be cured by expression of specific genes in a variety of cells. Blood-related deficiencies, such as the lack of clotting factors in hemophilia, can potentially be cured by enabling the cells lining blood vessels to produce the necessary clotting factors. A caution for untargeted therapy is that certain cell types may suffer complications from expression of inappropriate genes. Another potential complication is that a patient's immune system may reject an introduced gene product and the cells producing it, leading to destruction of healthy tissue.

Regulation of quantity and duration of gene expression is also important. While diseases such as cystic fibrosis and hemophilia require persistent expression and may be cured even with low expression, other diseases such as diabetes require tightly regulated and coordinated gene expression. For some genetic diseases, irreparable damage occurs early in life. For example, cystic fibrosis leads to lung damage during childhood. It is important to intervene and treat such diseases before permanent damage is sustained.

An early success for gene therapy was witnessed in a 1990 trial when two girls with a genetic deficit causing severe immunodeficiency were given infusions of their own immune system cells. These cells were genetically engineered to contain a working version of their missing gene. Following regular monthly administration, the girls developed active immune systems that allowed them to remain healthy for more than 10 years.

The first marketed gene therapy product was approved in China in 2003. Shenzen SiBiono's Genicide is targeted at head and neck squamous cell carcinoma, a highly lethal cancer with an annual incidence of 300,000 people in China. The drug uses a benign viral vector to deliver *p53*, a gene implicated in controlling cell growth; many tumors contain defective *p53* or fail to express sufficient quantities of the protein.

Despite its early promise, advances in gene therapy have been hampered by variability in the safety and effectiveness of trials, sometimes with fatal consequences.

DIAGNOSTIC TESTS

In addition to treating diseases, biotechnology has also made it easier to detect and diagnose medical conditions. A quantum leap past traditional techniques that require correlation of numerous symptoms to develop a diagnosis, biotechnology enables the direct detection of biological processes. In addition to refining symptom-based diagnoses, it is also possible to make determinations at earlier stages. Screening for pregnancy and cancer are examples of diagnoses that have increased in reliability and sensitivity as a result of biotechnology.

For diseases where symptoms are usually noticed past the point where treatment is most effective, diagnostic tests can save lives by enabling at-risk individuals to monitor their health prior to onset of disease. Individuals with genetic predispositions to specific cancers can be alerted to their increased likelihood of disease and can engage in preventative activities and regular screenings, potentially avoiding disease progression or allowing early intervention.

A secondary benefit of diagnostic tests is that they can enable individuals to avoid costly, dangerous, or unnecessary procedures (see Box *Personalized medicine and drug sales* later in this chapter). For example, use of aspirin to prevent colorectal cancer may be less cost-effective than regular screening. Citing a cost of nearly $150,000 per year of life saved by preventative use of aspirin, factoring in the costs of the drug and treatment of side effects, versus a $30,000 cost of screening per year of life saved, a recent study concluded that screening was preferable to aspirin use.[3]

On initial examination, diagnostics may seem to be a good market-entry objective for start-ups seeking to develop an ini-

3 Ladabaum, U., *et al.*, Aspirin as an adjunct to screening for prevention of sporadic colorectal cancer. *Annals of Internal Medicine*, 2001. 135(9):769-781

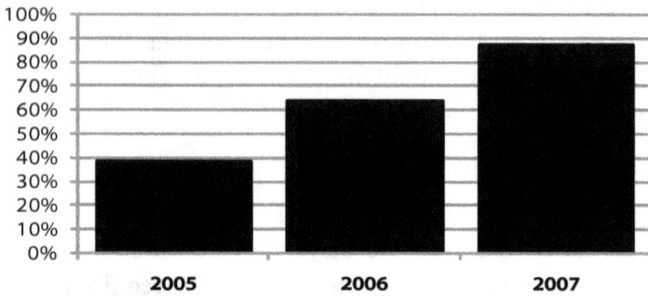

Figure 6.3 *U.S. babies born in states mandating genetic disorder testing*
Source: March of Dimes

tial revenue stream. Diagnostics are relatively cheaper to develop than therapeutics and it is also relatively easier to gain regulatory approval for them. However, because drugs serve more pressing needs than diagnostics they can be assured of relatively greater sales and greater profits due to decreased price elasticity.

PERSONAL GENETIC PROFILING

As information about genetic markers increases and DNA profiling and sequencing costs decrease, numerous companies are entering the personal genetic profiling space. These companies offer to profile or sequence an individual's DNA and identify predispositions to disease, with the objective of enabling their customers to take proactive approaches to prevent disease onset.

While the potential to use genetic information to help prevent disease is an exciting prospect, the science behind predicting disease predispositions is too immature to be useful in most cases. A federal panel investigating personal genetic tests concluded that "a growing number of the tests are being marketed with claims that are unproved, ambiguous, false or misleading" and added that a potential for harm exists "if a test is inaccurate, patients may be given risky, unnecessary treatments or denied treatments that would be highly beneficial."[4]

Aside from fraudulent operations seeking to sell vitamin mixtures of dubious benefit to consumers, the very science of disease predisposition mapping is in its infancy. Many diseases have strong environmental and genetic components, and the influences of these factors need to be understood for genetic profiling to be useful. Faced with a poorly-documented association of a genetic sequence with a disease, or conflicting reports of genetic predisposition, testing companies and patients are likely to assume a stronger connection than is warranted by the data.

PERSONALIZED MEDICINE

Personalized medicine involves the application of technologies such as functional genomics (see Chapter 5) to tailor therapies to the patients most likely to benefit from them. It is estimated that most commonly used drugs are effective in only 30–60 percent of patients with a given disease. A subset of these patients may suffer severe side effects. There are two causes for this difference in response. First, most diseases have myriad causes. They tend to be defined by symptoms, but each distinct cause may respond best to a different treatment. Second, people are different. Differences in liver metabolic enzymes, for example, can determine if patients are unlikely to respond to a drug or if they will suffer severe side effects—see the section *Pharmacogenetics* in Chapter 5 for more details.

Aligning treatments with the patients most likely to benefit from them holds the potential to improve the effectiveness of medical intervention, while reducing healthcare costs and dangers. The key to realizing personalized medicine is alignment of a diagnostic test with a therapeutic intervention. A well-paired test, such as the test for Her-2 overexpression which indicates that Genentech's Herceptin is the preferred drug, can have a strong positive impact on sales. In other cases, personalized medicine may mean fewer sales.

4 Pear, R. Growth of genetic tests concerns federal panel. *New York Times*, January 18, 2008.

Box
Personalized medicine and drug sales

Personalized medicine has great potential to improve the safety and efficacy of drugs by targeting therapies to those most likely to benefit from them and excluding patients who are susceptible to deleterious side effects. While patients generally stand to benefit from personalized medicine, biotechnology companies may benefit or suffer based on whether screening methods are used to identify, or to exclude, patients. The contrasting examples of Herceptin and Aczone demonstrate how personalized medicine can benefit or hurt drug sales.

HERCEPTIN

Genentech's Herceptin is a monoclonal antibody directed at the Her-2 cell receptor. Overexpression of Her-2 is implicated in approximately 20 percent of breast cancers. It is estimated that without a test for Her-2 overexpression, Genentech would have needed to perform clinical trials on 2,200 patients for ten years in order to demonstrate the efficacy of Herceptin.[1] Utilizing a test for over-expression of Her-2 to segregate patients, Genentech was able to demonstrate Herceptin's ability to safely increase survival times by 50 percent using only 469 patients in less than two years.

Because the test for Her-2 overexpression is tied to the mode of action of Herceptin, patients are able to avoid many of the unnecessary side effects associated with ineffective medicines and may benefit from early prescription of the drug most likely to effectively treat their tumors, while Genentech benefits from preferred prescription to patients who test positive for Her-2 overexpression.

ACZONE

QLT's Aczone is a topical drug used to treat acne. In the course of clinical trials it was found that people with a blood disorder called G6PD deficiency have a higher risk of developing anemia with Aczone; roughly 1.4 percent of patients in clinical trials had this disorder. The potential for anemia among patients with G6PD deficiency taking Aczone spurred the FDA to require that patients be tested for the enzyme deficiency prior to being prescribed Aczone.

Unlike Herceptin, where the diagnostic test is optional prior to prescription and identifies the target population, Aczone prescription requires prior testing to exclude patients likely to suffer deleterious side effects. The impact of this excluding screen is that while patients are protected from adverse reactions, QLT suffers a significant barrier to prescription.

1 Tansey, B. Genentech a big believer in diagnostics, *San Francisco Chronicle*, May 17, 2004. p. B-1.

While fewer sales may represent a barrier to implementation of personalized medicine in some cases, it can also be a benefit. Using profiling in clinical trials can speed approval. It is estimated that the time saved in Herceptin's clinical trials netted Genentech between $1.2 and $1.5 billion. Selling drugs to large populations also greatly increases the likelihood that significant side effects will emerge and may potentially lead to market withdrawal and debilitating lawsuits. The example of Vioxx (see Box *Vioxx: Anticipating and disclosing side effects* in Chapter 8) illustrates the risks encountered in serving large populations.

TISSUE ENGINEERING

Tissue engineering is the production of natural or synthetic organs and tissues which may be implanted as fully functional units, or as tissue which undergoes further development following implantation to perform necessary functions. The first engineered tissues were skin equivalents used to treat burn victims, and structural scaffolds to replace heart valves, arteries, and bones. Alternative treatments for tissue and organ failure include transplantation from donors, surgical repair, artificial prostheses, mechanical devices, and in a few cases, drug therapy. Tissue engineering has the potential to provide an alternative or complement to these treatments, potentially with fewer side effects and a greater ability to treat major damage.

While some cell types and tissues are amenable to production in liquid media or on solid surfaces, a challenge for production of more complex tissues and organs is the development of appropriate scaffolds to model growth and methods to direct local differentiation of tissues. Large organs must be perfused by blood vessels to allow for oxygenation, delivery of nutrients, and removal of waste. An alternative to laboratory production of implantable organs is the use of stem cells, which can potentially be coaxed to repair tissues upon injection into patients.

Tissue engineering also faces significant commercial challenges. Stem cells and xenotransplantation offer alternative methods to serve many of the same markets as tissue engineering. The history of tissue engineering also serves as an example of how quickly fortunes can turn in biotechnology. Despite research and development expenditures of $4.5 billion, as of 2002 none of the tissue engineering products on the market were profitable.[5] By 2007, the sector consisted of 50 profitable firms which had treated over a million patients and were generating $1.3 billion in annual sales.[6]

STEM CELLS

Stem cells can repair damaged organs and tissues and even have the potential to produce entire organs in laboratory settings for use as human replacement parts. Understanding and controlling the ability of stem cells to repair organs holds the potential to eliminate the need for transplantation and tissue engineering. Unlike most of the cells in human adults, stem cells are able to differentiate into other cell types. Degenerative diseases such as Alzheimer's disease and Parkinson's disease, as well as diseases marked by cell injury or malfunction, such as stroke, heart attack, cancer, and spinal cord injury, are all candidates for stem cell therapy.

Stem cells can be harvested from remnants of fertility treatments, placentas, umbilical cords, and from adult tissues. Stem cells from embryonic sources such as discarded fertilized eggs from fertility treatments carry a significant ethical and political burden. Additionally, there are restrictions on the use of federal funding for embryonic stem cell research in the United States. Because of the central importance of federal funding in supporting basic research (see Figure 4.3) and the potential for political sentiments to translate to regulatory decisions, the potential of embryonic stem cell research in the United States is uncertain.

---

5 Lysaght, M.J., Hazlehurst, A.L. Tissue engineering: the end of the beginning. *Tissue Engineering*, 2004. 10(2):309-320.

6 Lysaght, M.J., Jaklenec, A., Deweerd, E. Great expectations: Private sector activity in tissue engineering, regenerative medicine, and stem cell therapeutics. *Tissue Engineering Part A*, February 1, 2008. 14(2):305-315.

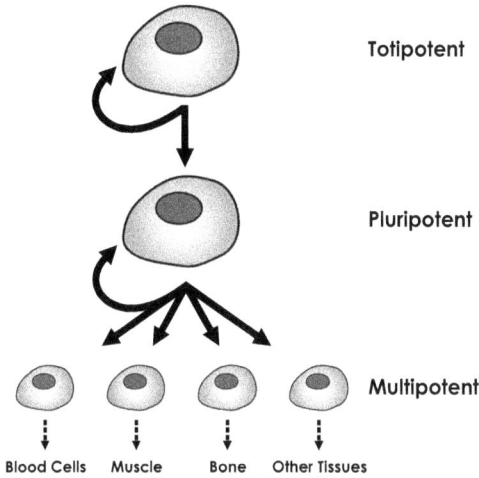

Figure 6.4 *Stem cell types*
Modified from National Institutes of Health

Stem cells extracted from adult sources such as bone mar-
row do not share the same ethical and funding issues as embryonic stem cells, but it is clear that they have different properties
as well. There are three categories of stem cells, distinguished
by their ability to differentiate into other cell types. Totipotent
cells can grow into an entire organism. Embryonic stem cells
are unique in having this property. Pluripotent cells cannot
grow into a whole organism, but they are able to differentiate
into the various cell types of the body and potentially form organs. Multipotent (sometimes called unipotent) cells can only
form certain types of cells such as blood cells or bone cells. Embryonic stem cells are pluripotent and multipotent by virtue of
their totipentcy, while adult stem cells may be pluripotent or
multipotent depending on their source.

A market for leukemia treatment with adult stem cells already exists. A growing body of research also indicates that
adult stem cells can cure hearts and other organs that have suf-
fered trauma. Interestingly, implantation of embryonic stem
cells has been implicated in cardiac arrhythmias and cancer development, suggesting that adult stem cells may be better suited

for these applications.

Stem cells are also being explored as a vehicle to deliver genes to specific tissues in the body for gene therapy or cancer treatment. The greatest technical challenge facing stem cell researchers is elucidating the factors that activate stem cells to form specific kinds of tissue. Isolating and purifying sufficient quantities of stem cells for clinical use presents an additional challenge. Extensive patenting also makes freedom to operate an important consideration.

XENOTRANSPLANTATION

Shortages of human organs available for transplant and disease considerations have prompted the search for alternative sources of organs. Xenotransplantation is the transplantation of organs from any other species into humans. The similarity of human and pig organs has led most xenotransplantation research to be focused on pigs. Immune rejection and the potential for novel infections to spread into human populations are the most pressing challenges facing xenotransplantation.

Xenotransplantation need not involve implantation of whole organs. Studies have shown survival of fetal pig neural cells when administered to patients with Huntington's or Parkinson's disease. The potential also exists to coax such cells to perform needed functions. It is also possible to use animal organs for *ex vivo* treatments; use without implantation. In 1997, a patient with liver failure had his blood perfused through a liver from a transgenic pig raised by Nextran (Baxter donated Nextran to the Mayo Clinic in 2003), keeping him alive for over six hours until a liver donor could be found.[7]

Rejection may occur by multiple mechanisms with fast or slow time profiles. Methods for dealing with rejection range from administering immunosuppressive agents to removing immunoreactive elements through genetic engineering of donor animals. Aside from rejection, a significant challenge for xenotransplantation is the prospect of novel infections being

7   Stolberg, S.G. Could this pig save your life? *New York Times,* October 3, 1999.

introduced through the transplant recipient into the human population. One way to partially alleviate this risk is to genetically engineer animals to remove known risk factors.

It is important to note that there are alternatives to xenotransplantation that avoid immune system rejection and disease issues. Tissue engineering and stem cells can potentially permit the growth of organs and tissues outside or inside the human body, eliminating the need for xenotransplantation.

III

Laws and Regulations

Product development in biotechnology relies on innovative research and development that expands the realm of scientific knowledge. The innovative nature of biotechnology research and the potential to do harm place significant burdens on biotechnology development. Unique legal, regulatory, and political issues differentiate biotechnology from other businesses. For example, new product development is expensive and time consuming whereas in many cases copying existing products is relatively straightforward. In order to support innovation and enable companies to recoup their research and development investments, intellectual property protection is required to maintain a competitive advantage.

Because many biotechnology products are based on existing life forms and may be intended for medical use by humans or exposed to the environment, it is very important to ensure that applications of biotechnology will not do harm. Government agencies such as the Environmental Protection Agency, Department of Agriculture, and Food and Drug Administration oversee biotechnology product development. Genetically modified microorganisms, plants, and animals must be proven to be environmentally safe before they can have an opportunity to interact with the environment. Drugs meant for human use must be thoroughly tested for safety and efficacy in humans before they can be granted marketing clearance.

Because biotechnology involves manipulations of living things, political and ethical decisions have an important impact on the ability to perform research and commercialize certain inventions. Bans on technologies such as cloning, stem cell research funding restrictions, and boycotts of genetically modified food can eliminate markets and influence the scope of scientific research.

Chapter 7

Intellectual Property

> Trying to patent a human gene is like trying to patent a
> tree. You can patent a table that you build from a tree,
> but you cannot patent the tree itself.
> *William Haseltine, former Chairman and CEO, Human*
> *Genome Sciences*

Intellectual property protection is essential in biotechnology because while the cost of innovation is high, the cost of imitation is relatively low. Unlike commodity-based industries, where access to cheap materials, labor, or markets can provide a competitive advantage, innovation-based industries such as biotechnology rely on the ability to generate and exploit knowledge to gain a competitive advantage. Research and development must be financed by sales, which can only occur after research and development have been completed. Intellectual property protection is necessary to secure a competitive advantage and ultimately promote innovation, enabling innovators to prevent competitors from offering prices that reflect their reduced R&D burden, Intellectual property protection therefore plays an integral role in enabling biotechnology research by establishing a barrier to competition that permits pioneers to sustain lengthy research efforts and recoup their research and

Table 7.1 *Intellectual property rights*

Patent	Prevent others from practicing an invention
Trade secret	Protect information and know-how
Trademark	Protect company and product name, look, and feel

development costs.

Intellectual property differs from other forms of property because it is the product of intellectual effort and may be embodied in a concept rather than a physical representation. There are three types of intellectual property rights relevant to biotechnology: patent, trade secret, and trademark. Patents allow an inventor to prevent others from practicing an invention without permission. Trade secrets are unique among intellectual property rights in that they protect information and know-how that are not in the public domain. Trademarks are words, symbols, or phrases used to identify a particular manufacturer or seller and their products and to distinguish them from others.

PATENTS

U.S. patent law grants the right to exclude others from making, using, offering for sale, or selling an invention in the U.S. Patent laws vary in different countries. Independent patent applications must be filed for each country in which protection is desired, although regional patent offices such as the European Patent Office enable patent protection in multiple countries. The commentary in this section is specific to the U.S. patent system, with important differences in other countries highlighted as necessary.

It is important to understand that patents do not grant the right to practice an invention, only the ability to exclude others from doing so. Concerns over patent holders "owning" genes arise from a misunderstanding; patents only grant the right to exclude others from novel applications of genes, patents do not assign ownership or the right to use a gene. Furthermore, patents cannot be used to protect naturally-occurring processes.

The objective of patent grants is to provide an incentive for innovation, and to reward inventors for publishing the best method to practice an invention. In return for disclosing the best mode for practicing an invention, inventors are given protection from competition for a number of years. In indus-

Table 7.2 *Top biotechnology patent holders*

Company	Number of patents (1976-2005)	2007 revenues ($millions)
Monsanto	3,763	8,563
Genentech	983	11,724
Chiron	834	1,920[1]
Amgen	699	14,771
Isis	698	70
Incyte	567	34
Human Genome Sciences	458	42
Millenium	424	528
ZymoGenetics	345	38
Genencor	320	410[2]

1. 2005 revenues; Chiron was acquired by Novartis in 2006.
2. 2004 revenues; Genencor was acquired by Danisco in 2005.

Source: Aggarwal, S., Gupta, V., Bagchi-Sen, S. Insights into US public biotech sector using patenting trends. *Nature Biotechnology, 2006.* 24(6):643-651.

tries such as biotechnology where reverse-engineering if often relatively simple, patents are essential to motivate innovation. Without patent protection, innovation of products subject to reverse-engineering would be discouraged by the inability to recoup research and development investments.

Patents filed prior to June 8th 1995 last the longer of 17 years from the date of issue or 20 years from the date of filing. The term of plant and utility patents filed after June 8th 1995 is 20 years from the date of filing.

When a patent expires, the subject matter of the patent becomes part of the public domain. This is one of the sacrifices that are made in exchange for patent protection. An alternative to patent protection is to maintain an invention as a trade secret. While trade secrets do not have expiration dates, they offer no protection from competitors who may reverse-engineer or independently develop an invention; the potential to retain exclusive use of an invention for a longer term than granted by patent protection is countered by the potential for reverse-engineering and independent invention. The decision to patent an invention or maintain it as a trade secret requires consideration of the ease of reverse-engineering and independent discovery versus the benefits associated with the right to exclude others from use of the invention.

PATENTING BIOTECHNOLOGY

A case that arguably permitted the development of the biotechnology industry in the United States was settled in 1980 when the Supreme Court reaffirmed in *Diamond v. Chakrabarty* that genetically modified organisms were patentable. Working as a General Electric research scientist, Ananda Chakrabarty combined the individual genes that conferred the ability to break down discrete components of crude oil, all within a single bacterial strain. The original patent application consisted of 36 claims to genetically engineered bacteria containing elements enabling them to degrade crude oil, permitting application to cleaning up oil spills.

The examiner of Chakrabarty's patent allowed claims detailing the production and preparation of the bacteria, but did not permit claims to the bacteria themselves, claiming "as living things, microbes are not patentable subject matter." The Supreme Court disagreed, referring to the Patent Act of 1952 in which Congress intended statutory subject matter to include "anything under the sun that is made by man." Accordingly, Chakrabarty's oil-eating bacteria, as the product of human manipulations, were deemed eligible for patent protection.

WHAT IS PATENTABLE?

Biotechnology inventions may claim a novel product—a new "manufacture" or a new "composition of matter"—as well as applications of the resulting materials in therapeutics or diagnostics. Raw products of nature are not patentable. They are not novel. However, a new use for a known product may be claimed in a patent. It is also possible to patent novel and useful processes, or improvements on existing processes. Process claims, like product claims, cannot refer to naturally occurring instances. Furthermore, a synthetic composition is not patentable if it is identical to naturally occurring products.

Therefore, while the discovery of a newly identified product of nature is not patentable, the extraction or isolation of substances from their natural environment, to make them available in a useful form for the first time, is patentable. This allowance is the basis of many "gene" or "naturally-occurring" drug patents. The gene or naturally occurring substance itself is not the subject of the patent; the processes for purification and alterations are the protected elements. Patents on genes generally claim either recombinant or purified and isolated forms of a gene, or applications of a gene. The same holds for drugs which are based on biological extracts. One cannot patent such a drug, but it may be possible to patent the purification, delivery method, or novel applications of the drug.

Patents must also define a specific application. It is not sufficient to say that a patent can be used in a broad array of situations. Applicants are required to demonstrate the diversity of applications. In the 1999 case of *Enzo Biochem v. Calgene Inc.*, Enzo's patent asserted that their antisense technology was "broadly applicable with respect to any organism containing genetic material." The court ruled against Enzo after determining that they did not sufficiently provide direction or examples of how to practice the patented claims in diverse cell types.

PATENT REQUIREMENTS

There are three basic requirements for utility patents: nonobviousness, novelty, and utility. The criteria to satisfy non-obviousness and novelty requirements are dynamic, reflecting the expansion of knowledge in pertinent fields and further complicating the establishment and defense of patents. Patent laws are also subject to change, expanding or reducing the scope or terms of patents. The requirement for patent utility claims to be substantial, in addition to the requirements of specificity and credibility, was recently added to ensure that claims are in context with the nature of an invention.

Non-obviousness is judged by taking the frame of mind of an average person in a given field, with knowledge of all prior art. Prior art is the public knowledge that exists in a field. The existence of any prior art demonstrating that an invention is either not new or is obvious makes the invention ineligible for patent protection. Generally, if something provides new and unexpected results, it may be patentable. Novelty simply implies that something which is already known or patented cannot be patented.

In order to be deemed useful, a patent must describe a substantial, specific, and credible application. Patents do not apply to theoretical phenomena or ideas. The requirement for claims to be substantial means that it is not possible to patent something that is potentially useful but has not been fully investigated. The utility must define a real-world context of use and must be consistent with the properties of an invention. One cannot patent a potential anti-cancer drug for use as landfill or a genetically engineered mouse as snake food in order to avoid demonstrating utility.

WHEN PATENTS EXPIRE

The price of patent protection is disclosure. The best mode of practicing the invention must be disclosed in exchange for the right to exclude others from using an invention. When a patent expires, the ability to practice the invention is no longer controlled by the inventor, so disclosure can facilitate the development of competing products following patent expiration. For products like drugs, pioneers can lose a significant portion of market share overnight with the introduction of generic versions. Each day that competition can be delayed may be worth millions of dollars in revenues.

Whereas the development of new drugs requires years of research to identify and refine potential drugs and determine their safety and efficacy, generic drug development generally requires less research and development. Because generic manufacturers can avoid much of the research and development expense, they can charge significantly less than pioneers and still earn a profit. Furthermore, unlike pioneers who may only be able to make rough estimates of the potential market for a drug prior to launch, generic manufacturers can look to pioneer sales to determine market size. Generic manufacturers therefore benefit from lower development risks and relatively greater certainty of revenues (generic drugs are described in greater detail in Chapter 8).

EXTENDING PATENT PROTECTION

Because of the ability of patents to protect markets, much effort is dedicated to ensuring that patent protection is extended as long as possible. Common methods to extend patent life are to utilize market exclusivity protections granted by the FDA, and to use R&D to leverage brand strength.

FDA market exclusivity creates incentives for socially-beneficial activities which companies might otherwise not pursue. Perhaps the most popular of these incentives is the Orphan Drug Act, which provides temporary market exclusivity for drugs addressing diseases affecting small populations. The purpose of this exclusivity is to motivate the development of drugs for markets which might not otherwise provide sufficient financial merit. Additionally, companies responding to FDA requests to conduct clinical trials demonstrating that drugs are safe and effective for pediatric patients or to provide supportive evidence demonstrating that drugs can be responsibly sold and used over the counter, called an OTC-switch, are also granted temporary market exclusivity. FDA market exclusivities are described further in Chapter 8.

Another strategy to prevent generic competitors from gaining market share involves a combination of R&D and marketing. New patents can be acquired by developing new variations of drugs such as new formulations, novel combinations, new delivery methods, and gaining approval for new indications. The repeated practice of launching these line extensions is called evergreening. Because evergreening does not prohibit generic sales of drugs with expired patents, success is contingent upon consumers being willing to pay a premium for the new product rather than accepting a less expensive generic version of the original product. Some companies specialize in developing and patenting improved formulations of drugs facing expiration and selling the rights back to pioneers to give them leverage against generic competitors. R&D strategies to leverage existing products are described in Chapter 11.

PATENT CHALLENGE

There are a variety of oversights that can lead to a patent being deemed invalid or unenforceable. The revelation of a prior publication that renders a patent obvious or not novel can lead to invalidation. Falsification of data or failure to disclose the best method to practice a patent can also invalidate a patent. An inventor who has not assigned his/her rights to an invention may practice the patented invention and may grant licenses to third parties without the permission of the co-inventors. The sponsoring institution that funded research leading to a patent may also have claim to partial ownership of an invention. In addition, all inventors must usually join as plaintiffs in an infringement suit. By being excluded from an infringement suit, a co-inventor can prevent the enforcement of a patent. Therefore, failure to list all the inventors of a patent can frustrate licensing agreements and potentially render a patent unenforceable.

While the U.S. Patent and Trademark Office is responsible for granting patents, the courts have the final determination of patent validity. Patents can be invalidated based on technical or legal grounds. Proof of failure to satisfy the requirements of novelty, nonobviousness, or utility can render a patent invalid. A downside of resorting to patent challenge is that biotechnology-related patent litigation is very expensive, typically costing $3-10 million for each party.[1]

---

1 Apple, T. The coming US patent opposition. *Nature Biotechnology*, 2005. 23(2):245-247.

TRADE SECRETS

An alternative to patenting inventions is to retain them as trade secrets. A trade secret is an item of information—a customer list, business plan, or manufacturing process—that has commercial value and is not exposed to the public by the holder. There are two basic types of trade secrets. Technical secrets refer to research and development methods and tools; business secrets refer to items such as marketing, sales, financial, and administrative data. A distinction is also made between an employee's abilities and a company's proprietary knowledge. A scientist's research expertise does not qualify as a trade secret, but a company's unique research techniques and methods may qualify.

Unlike the limited term of patent protection, trade secrets can potentially last indefinitely. A limitation of this indefinite term is that the holder of a trade secret has a responsibility to make efforts to protect it. Courts may refuse to recognize a trade secret if reasonable efforts have not been made to keep the information from being disclosed.

A disadvantage of trade secrets, when compared to patents, is that they do not grant the right to exclude others from practicing an invention. If a competitor is able to independently develop or reverse-engineer an invention, or if the owner of the secret accidentally makes it public, then the owner has little recourse for preventing use.

TRADEMARK

A trademark is a word, symbol, or phrase used to identify a particular manufacturer or seller and their products, and to distinguish them from others. Unlike copyrights or patents, trademark rights can last indefinitely if the owner continues to use the mark to identify its goods or services. Trademarks can also extend to the look and feel of a product or logo, provided it does not confer any sort of functional or competitive advantage. Drug companies trademark the names and physical characteristics of their drugs, which is why generic drugs have different names and look different from pioneer products.

The rights to a trademark can be lost through abandonment, improper licensing or assignment, or genericity. Non-use of a trademark for three consecutive years is considered evidence of abandonment. One reason why companies are very particular about the representation of their trademarks is that a trademark licensed without adequate quality control or supervision by the owner may be canceled. The rationalization in such situations is that the trademark no longer identifies the goods of a particular provider.

A relevant example of the loss of a trademark to genericity is Bayer's loss of its Aspirin trademark. Because the word *aspirin* was consistently used without the product descriptor *pain reliever* following it, aspirin became a generic term for pain reliever and was therefore no longer protected under trademark law.

THE ROLE OF INTELLECTUAL PROPERTY IN BIOTECHNOLOGY

This chapter opened with a quotation that addresses a common misconception in patenting biotechnology inventions. It is not possible to patent genes; only applications of genes may be patented. Biotechnology research is focused on discovering and enabling applications of molecular biology. The emphasis on developing and commercializing novel products and services means that intellectual property protection is essential to foster innovation in the biotechnology industry. Patents and other forms of intellectual property enable companies to attract financing and endure lengthy development efforts by providing market exclusivity for commercial applications.

The benefits of market exclusivity are countered by concerns that patents impede scientific progress and limit access to useful tools and products. While intellectual property protection does enable companies to ask relatively higher prices for protected products, it is important to consider the basis of intellectual property protection.

Ideally, the purpose of intellectual property protection is to enable and accelerate development of inventions that would not occur without protection. Substantial investments are required to produce innovative products and services. In drug development, for example, the bulk of effort is spent identifying potential drugs and proving them to be safe and effective. Many years and hundreds of millions of dollars are required to develop and demonstrate the safety and efficacy of even a single drug.

The sophistication of research tools means that it is often possible to reverse-engineer a proven drug or other biotechnology-derived product. Without patent protection, competitors would be free to copy successful products at a much lower development cost than pioneers, and could therefore profitably sell them at lower prices. This premature price competition would prevent pioneers from recouping their research and development investments, favoring a commodity-based rather than an innovation-based industry model. This is exactly what was witnessed in China and India before their patent laws were strengthened to protect drugs; these countries had thriving generic drug industries with strong exports, but little inward investment or research and development activity.

To avoid the unwelcome situation in which competitors prevent pioneers from securing compensation for their efforts, research is focused on applications that can be protected by patents or other means. The scope of what can be protected thereby indirectly defines which applications biotechnology companies will develop.

In addition to protecting intellectual property and establishing a virtual monopoly, patents also have some indirect benefits. During development, in the absence of profits, patents can imply the ability to generate profits. The demonstrated ability to turn ideas into inventions can attract investors and development partners. Revenues can also be derived from the sale or license of patents.

Concerns over restriction of research due to an overwhelming number of patents are somewhat resolved by the decrease in patent value as the number of patents increases; the price of patents must reflect their value. A single high priced patent that controls an area of research may also encourage competitors to develop alternative methods and license them at lower prices. Patents, like physical property, are subject to the effects of supply, demand, and competition.

The requirements for novelty and non-obviousness dynamically reflect the current state of scientific knowledge. Furthermore, in order to be granted a patent, an inventor must disclose publicly how to best practice the invention. Therefore, patents provide a means by which the public can gain valuable cutting-edge scientific knowledge and abilities in exchange for a temporary grant of monopoly, which allows innovators to recoup their investments in research and development.

Chapter 8

Regulation

> Perhaps the most important discovery of the twentieth
> century was to learn to identify and read the code of life.
> And perhaps the greatest challenge we will face in the
> twenty-first century ... is how ... and when ... to apply
> this knowledge.
> *Juan Enriquez, Harvard Business School*

R egulatory approval is the second most important measure of the quality of a product from an investment perspective, after patent strength. Regulations ensure that
drugs are safe, effective, and appropriately labeled, and that biotechnology products such as plants and bioremediating bacteria
will not harm the environment. In addition to protecting public
health, regulations also enable objective measures of product
efficacy and safety, facilitating their comparison.

Regulations serve the dual functions of setting limits on
the applications of biotechnology and providing incentives for
innovation. An example of an regulation which drives innovation, the Orphan Drug Act, provides incentives for companies

Table 8.1 *Biotechnology regulating bodies*

Food and Drug Administration	Food, feed additives, veterinary drugs, human drugs, and medical devices
Department of Agriculture	Plant pests, plants, and veterinary biologics
Environmental Protection Agency	Microbial and plant pesticides of chemical and biological origin, new uses of existing pesticides, and novel organisms that may have industrial uses

to produce treatments for rare disorders. These incentives aim to encourage research into areas that might otherwise not demonstrate sufficient profit potential to merit investigation.

The agencies primarily responsible for regulating biotechnology in the United States are the Food and Drug Administration, Department of Agriculture, and Environmental Protection Agency. Products are regulated according to their intended use. Some products are regulated by more than one agency.

FOOD AND DRUG ADMINISTRATION

The FDA is responsible for regulating food, feed additives, veterinary drugs, human drugs and medical devices. Drug makers must test and gather data showing whether a drug is stable at certain temperatures and in powder, injectable, pill, or tablet form, and whether it can be manufactured repeatedly with consistent quality. In addition to defining a development path by which to demonstrate the safety and efficacy of drugs, clinical trials also play a vital role as valuation milestones.

The FDA also regulates manufacturing and delivery processes. Good manufacturing practice guidelines ensure the quality and purity of chemical products that are intended for use in pharmaceutical applications and describe the controls to ensure that the methods and facilities used for production, processing, packing, and storage result in drugs with consistent and sufficient quality, purity, and activity.

CLINICAL TRIALS

Biotechnology drug development is covered in detail in Chapter 4. The clinical trial process is described here. The basic process for drug development is as follows:

1. Potential drug compounds are identified, using laboratory and animal models.
2. Potential drugs are tested in animals and cell cultures to determine if they are safe enough for clinical trials in humans. These pre-clinical tests attempt to predict the ways

in which drugs may interact with the human body.

3. If a drug passes pre-clinical development, human clinical trials follow to ultimately provide information, necessary for FDA approval, on its safety and efficacy.

On average, one compound in a thousand will make it to clinical trials. Roughly 70 percent of drugs that complete clinical trials receive FDA approval.

Clinical trials are conducted to determine the safety and efficacy of drugs. Drugs must be proven safe and effective before FDA approval for marketing can be granted. While safety is the primary concern, a drug with detrimental side effects may be acceptable if there are no better treatments and the severity of disease warrants it. Most companies file for and receive patents for the commercial use of compounds during pre-clinical development. Much of a drug patent's life can therefore lapse during clinical trials and while waiting for regulatory review. The Hatch-Waxman Act, described later in this chapter, contains provisions to partially recover time spent in clinical trials and waiting for FDA approval.

Every clinical trial in the United States must be monitored by an Institutional Review Board (IRB), an independent committee of physicians, statisticians, community advocates, and others that ensures that the risks are as low as possible and are worth any potential benefits, and that the rights of study subjects are protected.

PHASE I

While the purpose of clinical trials is to determine the safety and efficacy of a drug, the primary consideration is the safety of the participants. Phase I trials are designed to determine the safety of drugs. These trials involve a small number of healthy volunteers or affected patients who are given doses ranging from sub-clinical to potentially toxic.

To minimize risk to human subjects, all drugs must undergo extensive pre-clinical development to determine their effects on animals, and predicted effects in humans, prior to Phase I trials.

Beginning with human trials in Phase I, drugs must be produced under current good manufacturing practices (cGMP). To satisfy FDA cGMP guidelines, manufacturers must be able to demonstrate compliance with regard to facilities, raw materials handling, and manufacturing control and associated documentation (see *Manufacturing* in Chapter 5).

There are two basic types of Phase I trials. *First-in-man* studies are primarily concerned with establishing the safety of a compound. These studies start with an initial small dose that is given to a small group of participants. If no adverse effects are seen, escalating doses are given to new groups of participants. Dose limiting toxicity is observed when the dose is escalated to the point that dangerous side effects are seen. The other type of Phase I study is the *clinical pharmacology* study. The goal of this study is to determine the pharmacokinetics of a compound: how a drug is absorbed, distributed in the body, metabolized, and excreted.

Data from Phase I trials are essential for the design of appropriate Phase II trials. Taking shortcuts in Phase I trials may only serve to see a compound fail in more expensive Phase II or Phase III trials.

Phase I testing ranges from one to three years on average. Historically, drugs in Phase I have a 10 percent chance of making it to market. If Phase I trials do not reveal unacceptable toxicity, a drug can proceed to Phase II testing. While failure in a phase I trial indicates that the tested form of a drug is unacceptable, success may still be possible by modifying a compound based on observed data.

PHASE II

The emphasis in Phase I trials is on safety; Phase II trials introduce effectiveness. Phase II trials consist of small, well-controlled experiments to further evaluate a drug's safety, assess side effects, and establish dosage guidelines. Drugs are given to volunteers (usually between 100 and 300 patients) who actually suffer from the disease or condition being targeted by a drug.

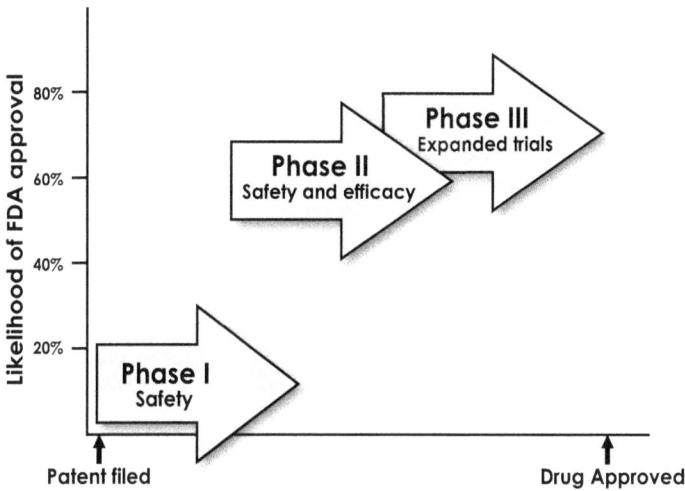

Figure 8.1 *Clinical trial phases*

This phase is where the minimum effective dose, maximum tolerable dose, and optimum dosage of drugs are established. Drug regimens are tested to see how often a drug must be administered; a drug may be effective if taken once a month, or may require administration several times a day. Statistical end points are established for drugs, representing the targeted favorable outcome of the study. The current standard of care for a medical condition can be used as a benchmark in setting the end point.

Phase II trials last an average of two years. If Phase II trials indicate effectiveness, a drug can proceed to Phase III trials. A drug that moves on to Phase III testing has an approximately 60 percent chance of being approved by the FDA. A properly designed and administered Phase II trial can help select dosage regimens and treatment indications that make Phase III trials faster and easier. Rushing this process may require repetition of Phase III trials or lead to outright failure.

Erbitux: Poor study design

ImClone's Erbitux originally failed to achieve a biologics license approval due to poor study design in a high-profile spectacle that saw ImClone's CEO Sam Waksal and media magnate Martha Stewart imprisoned amid allegations of insider trading. Erbitux was eventually approved, but had it failed outright it would have cost senior partner Bristol-Myers Squibb nearly $2 billion, making it an excellent case study of drug development gone wrong.

Erbitux is a monoclonal antibody that works by binding and inhibiting receptors that ordinarily signal cells to divide, preventing the proliferation of cancer cells. ImClone filed for an IND for Erbitux in 1994. ImClone researchers examined the safety and efficacy of the drug for a variety of cancers, including as a treatment in combination with a standard chemotherapy drug, irinotecan, in refractory colorectal cancer.

A key Phase II trial enlisted 139 patients and measured the effectiveness of Erbitux in combination with irinotecan in reducing tumor size—a surrogate endpoint for patient survival—and was intended to examine the effectiveness of the combination for third-line therapy for patients with metastatic cancer who had failed to respond to irinotecan alone. While this trial was not intended to be used as a pivotal trial for approval, encouraging results led ImClone's management to file a biologic license application (BLA) on Erbitux earlier than planned. In a meeting with the FDA, ImClone and the FDA agreed that in order for a BLA to be granted, (1) patients in the study must have tumors that progressed despite prior treatment with irinotecan; (2) at least 15 percent of the patients must respond to the combined regimen of Erbitux and irinotecan, with at least a 50 percent reduction in tumor size; and (3) the findings must meet statistical requirements.

In granting fast track status to Erbitux, the FDA required ImClone to conduct a small-scale study of Erbitux as a single agent (i.e., not in combination with irinotecan) for patients with colorectal cancer refractory to irinotecan, to set a baseline for the combination therapy and to demonstrate that irinotecan was essential for Erbitux's effectiveness.

Around this time, Bristol-Myers Squibb entered into a $2 billion agreement with ImClone, agreeing to purchase 19.9 percent of ImClone's stock for $1 billion (an approximately 75 percent premium over the market price) and offering an additional $1 billion in milestone payments connected to Erbitux development and commercialization.

While the combination of Erbitux and irinotecan demonstrated a 22.5 percent response rate (measured by tumor shrinkage) and the single agent Erbitux study showed a 10.5 percent response rate, the results of the two trials were not statistically distinguishable. The interpretation of this finding is that ImClone had failed to demonstrate that combination therapy with irinotecan was necessary for Erbitux to be effective. Furthermore, missing data documenting that patients in the combination study were refractory to irinotecan led the FDA to question whether it was Erbitux or irinotecan that was responsible for tumor shrinkage in that study.

Without strong data demonstrating that combination therapy with irinotecan was necessary or that Erbitux was effective against tumors refractory to irinotecan, the FDA rejected Erbitux's first BLA. Erbitux was eventually approved after performing a new study and recovering and submitting much of the missing data.

Phase IIA / IIB

Because clinical trial phases can take years to complete and often have multiple objectives, drug developers have taken to sub-dividing the phases in order to express a sense of progression. This division is most prevalent in Phase II trials, where safety data from Phase I trials are confirmed and expanded, and dosage and administration profiles are established in preparation for Phase III trials.

While the terms Phase IIa and Phase IIb are not recognized by the FDA, they are used useful devices to convey a drug's position in the approval process to investors, analysts, and partners. As a general rule, Phase IIa trials tend to address expansion and confirmation of data from Phase I trials—absorption, metabolism, and pharmacodynamics—whereas Phase IIb trials resemble small-scale Phase III trials in their evaluation of safety and clinical efficacy in large populations.

PHASE III

Phase III testing is the largest and most expensive clinical trial phase, and is intended to verify the effectiveness of a drug for the condition it targets, based on statistical end points established in Phase II trials. Phase III trials also continue to build the safety profile of drugs and record possible side effects and adverse reactions resulting from long-term use.

Phase III trials are tightly controlled, preferably double-blind, studies usually with at least 1,000 patients. In double-blind studies, neither patients nor the individuals treating them know whether the active drug or an alternative such as a placebo is being administered. Relative to Phase I and Phase II trials, the larger and ideally more diverse populations used in Phase III trials are necessary to determine if certain types of patients develop side effects or do not respond to treatment.

Two successful Phase III trials are generally required to ensure the validity of the studies, although a single trial may suffice if the results are extremely strong. Phase III testing averages between three and four years.

APPROVAL

Assuming a drug reaches the desirable end point in Phase III trials, the sponsor company will then file a new drug application (NDA) or biologic license application (BLA), which contains detailed information supporting the efficacy and safety of the drug. NDAs are submitted to the Center for Drug Evaluation and Research (CDER) and describe small molecule therapeutics that can be discretely defined. BLAs covering therapeutic applications such as protein-based drugs, growth factors, and antibodies are also submitted to CDER. BLAs covering other purified biological products such as blood products, vaccines, gene therapy vectors, and antitoxins are submitted to the Center for Biologics Evaluation and Research. The decision of which FDA center processes an application depends on the availability of appropriate expertise to evaluate a drug.

Following NDA/BLA submission, a drug has a better than 70 percent chance of being approved. At the FDA, a review team of medical doctors, chemists, statisticians, microbiologists, pharmacologists and other experts evaluates whether submitted studies demonstrate that a drug is safe and effective for its proposed use.

An application must provide sufficient information, data, and analysis to permit FDA reviewers to determine if a drug is safe and effective for the proposed use(s), if the benefits of the drug outweigh the risks, that the proposed labeling is appropriate, and if the methods used in manufacturing and the controls used to maintain quality are adequate to preserve the drug's identity, strength, quality, and purity.

Approval of an application can take anywhere from two months to an extreme of several years, if the FDA requests additional information. The Hatch-Waxman Act permits day-for-day recovery of patent life for time spent waiting for FDA approval (see *Hatch-Waxman Act* later in this chapter).

Following FDA approval, a company may market and distribute a drug to the patient population determined in Phase III trials. At this point, a drug is likely protected by a patent that extends 20 years from the date of patent application, which was sometime before clinical trials began. Post-approval patent life-spans often range from 8-12 years.

Phase IV

Once a drug is on the market, the sponsor must continue to perform observational studies in an ongoing evaluation of the drug's safety during routine use. Follow-up Phase III trials to confirm the safety and efficacy of drugs approved under accelerated development and review are sometimes referred to as Phase IV trials. In other cases, the safety and effectiveness of drugs may be monitored in applications other than those originally approved by the FDA.

Additional uses for many drugs are found after their initial launch. Accordingly, the objective of many companies is to find a good first indication for a drug in order to gain FDA approval. This demonstration that the drug has passed the FDA's safety assessment can be leveraged to gain marketing approval of additional indications for which a drug is effective.

An inherent risk of Phase IV trials is that they may reveal

new data on side-effects or dangers associated with a drug. Failure to quickly and transparently respond to such information can provide a base for future tort lawsuits.

Box

Vioxx: Anticipating and disclosing side effects

In September 2004 Merck voluntarily pulled Vioxx, a drug generating more than $2 billion in annual sales, from the market following evidence that the drug could be responsible for heart complications seen in long-term Vioxx users. In the following months thousands of lawsuits were filed against Merck, alleging that Merck failed to adequately warn patients and intentionally hid data on dangerous side effects for years.

When Vioxx and other so-called Cox-2 inhibitors were first discovered, they were hailed as next-generation anti-inflammatory drugs. Comparable first-generation anti-inflammatory drugs such as aspirin act by inhibiting both Cox-1 and Cox-2 enzymes. Inhibition of Cox-2 without affecting Cox-1 was hoped to reduce inflammation without the intestinal irritability associated with inhibiting both enzymes.

The mutually-opposing activities of Cox-1 and Cox-2 enzymes suggest the potential for vascular complications. Cox-1 and Cox-2 play counteracting roles in affecting narrowing of blood vessels and clot formation: inhibition of Cox-1 impedes blood vessel narrowing and clot formation, while inhibition of Cox-2 inhibits blood vessel dilation and enables clot formation.

Early trials in 2000 comparing Vioxx and naproxen, a non-specific Cox-1 and Cox-2 inhibitor, showed increased incidence of heart complications with Vioxx, but these were attributed to the possibility that naproxen offered greater protection against these complications. In 2002 the FDA requested a labeling change stating that these initial findings should be included on the label.

In a later trial investigating the ability of Vioxx to prevent the formation of colon polyps, the group receiving Vioxx experienced roughly twice the rate of serious cardiovascular side effects such as heart attacks and stroke as the untreated group. Unlike the aforementioned comparative study, this placebo-controlled study was able to discern the negative effects of Vioxx, illustrating the importance of using placebos in medical research. This finding led to the March 2004 withdrawal. Shortly thereafter, other manufacturers of Cox-2 inhibitors added warning labels to their drugs but declined to withdraw them.

Off-label use

While sponsoring companies are not allowed to advocate usage for non-FDA approved indications, physician-initiated "off-label" use can expand the market for a drug. For example, Cephalon's narcolepsy drug Provigil is largely used as a non-sedative treatment for depression.

A potential problem for popular drugs is that they may be prescribed to populations that are larger and more diverse than the clinical trial participants, and for conditions beyond the original indications. In some cases, the side effects that emerge can cause a drug to be shelved for all uses, underscoring the importance of monitoring off-label use.

Cephalon's management of off-label use of its cancer pain-killer Fentora is an example of proactive communication to control off-label use. After reports of four deaths related to the use of Fentora to treat conditions such as headache—not an approved indication for the drug—Cephalon sent letters to doctors reaffirming that Fentora was only to be used for cancer pain that is untreatable by regular doses of pain medicines.

Accelerated Approval

Accelerated approval makes promising products for life threatening diseases available on the market prior to formal demonstration of patient benefit. Whereas the traditional approval process requires that clinical benefit be shown before approval can be granted, accelerated approval allows a new drug application to be approved before measures of effectiveness that would usually be required are available. Surrogate endpoints—indirect measures of effectiveness such as laboratory findings—are used to show the strong potential for effectiveness in accelerated development and review. An important element of accelerated development and review is that testing must continue after the drug is approved to demonstrate its projected safety and efficacy. Failure to meet projected endpoints can result in withdrawal from the market.

FAST TRACK

The fast track designation is often confused with accelerated approval, but the two vary greatly in the circumstances under which they are granted and the regulatory processes they impact. Whereas accelerated approval makes experimental drugs available for life threatening diseases based on surrogate markers (indirect measures of efficacy such as tumor shrinkage), fast track is a process for interacting with the FDA during drug development. Fast track status grants drug developers scheduled development planning meetings with the FDA, the option of requesting the use of surrogate endpoints in evaluating studies rather than survival or other hard demonstrations of efficacy, and the option of submitting a BLA or NDA in sections on a rolling basis and authorizing the FDA to begin review of the application prior to its completion. Fast track status does not mean that a drug will receive faster approval, although priority review may be granted.

MARKET EXCLUSIVITY

In addition to market exclusivity derived from patents, the FDA also grants market exclusivity to drugs meeting special conditions. The motivation to grant exclusivity is to foster innovation and promote the development of drugs for applications that might otherwise offer insufficient motivation.

NEW DRUG PRODUCT

New drug product exclusivity (also known as new molecular entity protection), provided by the Federal Food, Drug, and Cosmetic Act, grants the holder of an approved NDA limited protection from new competition in the marketplace for the innovation represented by its approved drug product. A five-year period of exclusivity is granted to NDAs for new drug products—products containing chemical entities never previously FDA approved alone or in combination.

NEW CLINICAL INVESTIGATION

A 3-year period of exclusivity is granted for drug products containing previously approved active elements, when the application contains reports of new clinical investigations that were essential to approval of the application. For example, changes in an approved drug product that affect its active ingredient(s), strength, dosage form, route of administration or conditions of use may be granted exclusivity if clinical investigations were essential to approval of the application containing those changes.

PEDIATRIC USE

Pediatric use exclusivity is the only form of exclusivity that provides extensions which initiate at the termination of other exclusivity protection (e.g., new drug product or new clinical investigation) or patent protection. Sponsors that complete FDA-requested pediatric clinical investigations can be granted two separate six month extensions.

Companies do not need to gain, or even seek, new marketing approval to receive pediatric exclusivity extensions, they need only perform the requested studies. In 2007 AstraZeneca received a six month pediatric extension for its breast cancer drug Arimidex for investigating the therapeutic potential of Arimidex in pediatric conditions that result from increased estrogen production. AstraZeneca's studies reportedly did not show measurable benefits in treating the investigated conditions, leading them to forgo pursuing marketing approval for those indications.

OVER-THE-COUNTER

The political motivations to provide incentives for over-the-counter (OTC) drug sales without a prescription are to improve patient autonomy and accessibility to health care, and to reduce the cost of health care. These benefits must be weighed against hazards such as the potential for individuals to improperly self-diagnose their health conditions and needs, or the benefits of physician screening. Some conditions may be indicative of future health problems and while their direct treatment might not necessitate a physician visit, proactive screening for associated conditions can be beneficial.

The motivations for drug companies to apply for OTC approval for a prescription drug—called an OTC-switch—are manifold. First, increased patient access can dramatically improve sales volume. Second, if the FDA deems that additional trials are necessary to prove safety and efficacy sufficient for OTC administration, applicant firms can be granted a 3-year exclusivity for the OTC market. Finally, drug companies may initiate an OTC-switch knowing that if they do not, a generic company may initiate a switch and obtain exclusivity.

ORPHAN DRUGS

Individuals affected by rare disorders often require many tests to confirm diagnosis and often must travel great distances to reach specialists with necessary expertise to diagnose and possibly treat the disease. Accordingly, drugs for so-called orphan diseases can have a significant impact on the quality of life of affected individuals and their families.

The Orphan Drug Act, enacted in 1983, provides incentives for companies that develop treatments for conditions affecting fewer than 200,000 Americans, or those that affect more than 200,000 Americans for which there is no reasonable expectation that the cost of research and development will be recovered

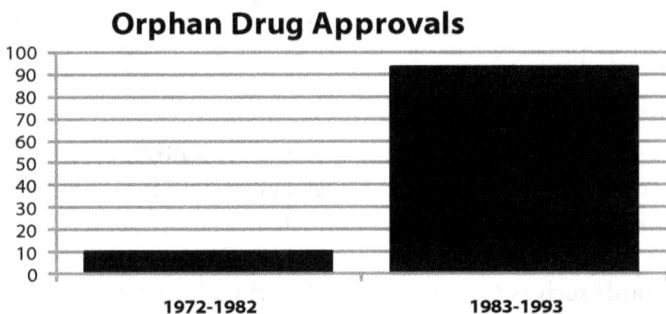

Figure 8.2 *Impact of the Orphan Drug Act*
Source: FDA

from U.S. sales. Companies are given clinical testing grants, tax credits for the costs of clinical trials, and seven years of market exclusivity for approved products, regardless of patent protection. Similar programs exist in Japan, Australia, and the European Union. The effectiveness of the Orphan Drug Act has been significant. Whereas 10 orphan drugs were approved in the decade preceding the Orphan Drug Act, 93 orphan drugs were approved in the decade following enactment (see Figure 8.2).

Despite being developed for small populations, not all or-

Box

Genzyme: Building an enterprise on orphans

Genzyme Corporation built a substantial portion of its enterprise by focusing on a therapeutic category which many other companies ignore: orphan drugs.

Genzyme's most lucrative niche is in lysosomal storage disorders. Cerezyme and Fabrazyme are enzyme-replacement therapies for Fabry disease, Aldurazyme is an enzyme-replacement therapy for Mucopolysaccharidosis I, and Myozyme is an enzyme-replacement therapy for Pompe disease. These treatments comprised nearly half of Genzyme's $3.8B revenues in 2007, with Cerezyme accounting for the largest share at $1.1B.

An issue facing Genzyme is that its drugs are very expensive. A year's dosage of these orphan drugs costs approximately $200,000. Careful management of drug pricing and access is crucial, as loss of insurance coverage, or loss of public and political support, could derail the company's business model.

Even with insurance coverage, the burden on patients can be significant. Genzyme defends its pricing by explaining that clinical trials for orphan drugs are very expensive, often requiring patients to be repeatedly flown to distant trial centers, and the drug products themselves are expensive to manufacture. Without a large population to generate sufficient revenues, the per-patient cost for orphan drugs must therefore be high. Genzyme also takes steps to ensure that no patient is denied access to their drugs. They offer numerous free drugs programs, covering patients in developing countries and those who are unable to access treatment despite living in countries with reimbursement programs.[1]

1 Free Drug Programs. Genzyme Corporation, 2008. http://www.genzyme.com/commitment/patients/free_programs.asp.

phan drugs serve small markets. In 2006, 19 orphan drugs generated revenues in excess of $1 billion. The high revenues from orphan drugs can derive from expansion into additional markets (either through off-label use or by gaining approval for new indications) or the ability to demand a high price due to a drug's efficacy and lower cost compared to alternative treatments.

GENERIC DRUGS

A generic drug is a version of a pioneer, or brand-name, drug that is produced by a second party. When the patent on a pioneer drug expires, competitors are free to market generic versions, provided they can receive FDA approval. Based on historical trends, generic drugs can be expected to capture as much as 60 percent of pioneer market share within one year of market entry. Furthermore, generic prices average about 61 percent of innovator drug prices in the first month of entry and drop to 37 percent within two years.[1]

To encourage the development of generic drugs, the Hatch-Waxman Act provides 180 days of generic market exclusivity to the first company that can gain approval for a generic drug. This exclusivity commences from the date of initiation of commercial marketing to permit time to amass sufficient stock for distribution. A 75 day deadline to initiate marketing was added in 2003 to prevent first-approval generic firms from indefinitely blocking others by postponing marketing.

The approval process for generic drugs is simpler than for pioneer drugs. Generic drug applications may avoid the lengthy and costly clinical trial process and can receive approval if they can be proven to be bioequivalent to a clinically proven pioneer drug.

TRADITIONAL PHARMACEUTICAL DRUGS

For the purposes of determining bioequivalence there are two categories of drugs: small-molecule synthetic drugs and biologics (see *Biotechnology vs. Traditional Pharmaceutical Drug*

---

1 Grabowski, H.G., Vernon, J.M. Effective patent life in pharmaceuticals. *International Journal of Technology Management*, 2000. 19(1/2):98-120.

Generic drug prices

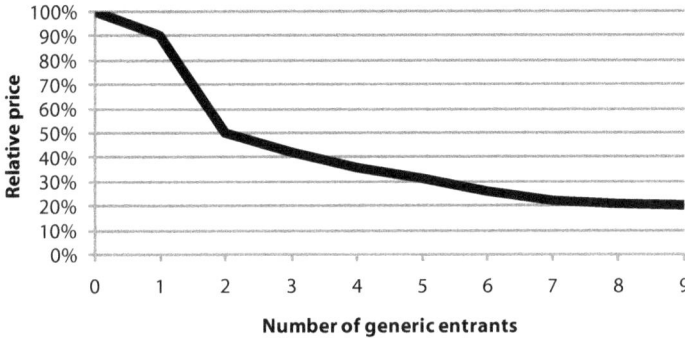

Figure 8.3 *Impact of generic entry on drug price*
Source: FDA analysis of IMS data from 1999 through 2004

Development in Chapter 4). Small-molecule synthetics are made using defined processes that result in a defined yield of product that is relatively pure. Biologics are derived from living sources and are much harder to objectively characterize. Examples of biologics include vaccines, gene therapy products, blood products, and tissues for transplantation. These preparations may be relatively unpure and the yield and precise nature of the product can also be difficult to define. The lack of consistency of biologic drugs does not imply that they are dangerous or ineffective; it just makes it more difficult to objectively compare pioneer and generic versions.

Generic manufacturers of small-molecule synthetic drugs must demonstrate identical chemical structure, purity, and concentration to the original drug. Generic drugs must also have the same pharmacokinetic profile as pioneer versions: the generic compound must be active at the same concentrations and for the same duration as the original compound. To gain regulatory approval for small-molecule synthetic drugs, generic manufacturers must file an abbreviated new drug application (ANDA) and use the same procedures, tools, and technologies as described in the original patents.

It can take two to five years to demonstrate bioequivalence and obtain approval for manufacturing facilities before an ANDA is granted. The production process and the source materials are strictly regulated. Just as two cooks following the same recipe may not produce identical products, generic manufacturers may face difficulty in producing a drug from the instructions in the original patent. In an interesting twist, generic producers may modify the production method and obtain patent protection for their improvements, potentially providing protection from other generics.

BIOLOGIC DRUGS

There is no established process by which generic biologics can be demonstrated to be identical to pioneer drugs. The lack of a means for generic manufacturers to leverage clinical trial data from pioneers raises the cost of market entry and in many cases enables pioneers to avoid price competition long after patents expire. The general model of generic biologic approval in the United States will likely be as "follow-on" drugs requiring limited clinical trials to demonstrate similarity to pioneer biologics. The impact of this system is that individual generic biologic drugs are likely to differ slightly from pioneer drugs and each other, potentially offering therapeutic improvements, opportunities for personalized medicine, and ultimately complicating prescription decisions.

In addition to uncertainty over the mechanisms of generic biologic approval and marketing, the expected cost savings are also in dispute. It is anticipated that generic biologics will have higher development costs, higher approval costs, and higher manufacturing costs than generic pharmaceutical drugs. If generic biologics have different efficacy and safety profiles than pioneer biologics, then they may end up with a smaller relative market share than generic pharmaceuticals are able to capture. The impact of these potential increased costs and decreased profits is that consumers will not realize the same savings with generic biologics that they have seen with generic pharmaceuticals. It is estimated that generic biologics will be sold at a 20 to 30 percent discount, versus generic pharmaceuticals which may be sold for discounts as high as 90 percent.

HATCH-WAXMAN ACT

The Hatch-Waxman Act, enacted in 1984, promotes innovation by fostering competition and allowing patent owners to recover time spent in clinical trials and FDA regulatory review. Measures easing the development of generic alternatives to pioneer drugs are balanced by patent term extensions that support pioneer drug manufacturers. The impact of the Hatch-Waxman Act has been significant. According to the Generic Pharmaceutical Association, the generic share of the prescription drug market has grown from 19 percent in 1984 to 53 percent in 2003, but accounts for only 12 percent of pharmaceutical costs.

Hatch-Waxman permits generic manufacturers to cite safety and effectiveness data from pioneer FDA applications, relieving the burden of performing lengthy and expensive clinical trials. To use pioneer safety and efficacy data, a generic drug must be demonstrated to be bioequivalent to a pioneer drug. In an exception to patent law, generic manufacturers are permitted to initiate clinical tests (necessary to demonstrate bioequivalence) before pioneer patents expire. As an incentive to promote the development of generic drugs, the first company to gain approval for the generic form of a drug is granted 180 days marketing exclusivity before other generic firms may enter the market.

To balance the benefits given to generic manufacturers, Hatch-Waxman also includes considerations for pioneers. A unique aspect of drugs, relative to most other marketable products, is that drugs cannot be marketed until they are proven safe and effective. This restriction places a significant financial and time burden on drug development firms, with a greater impact on innovative drugs requiring extensive clinical trials and possibly protracted FDA review. Hatch-Waxman grants an extra half-day restoration of patent life for every day of clinical trials and day-for-day restoration for the FDA review period. There are two restrictions on restoration: The effective life of a drug patent cannot exceed 14 years, and the total time restored cannot exceed five years.

DRUG NAMES

Drugs have several types of names, each of which is used for a different context (see Table 8.3). Chemical and biological names are respectively used to describe the composition of small-molecule or biological drugs. Generic names are shared between branded and generic drugs to indicate common ingredients. The trade name of a drug is the proprietary name used by different firms to brand their products.

The chemical or biological name of a drug is determined using several conventions, the objectives of which are to provide scientific descriptions of the composition of a drug. The chemical name for ibuprofen, for example, is *2-[4-(2-methylpropyl)phenyl]propanoic acid*; the biological name for Amgen's Epogen is *recombinant erythropoietin*. The generic name of a drug is created using a specific nomenclature system and is used to identify generic versions of a branded drug. The generic name for Amgen's Epogen is *epoetin alfa*. Ibuprofen is the generic name for a drug which has been marketed under Advil (Wyeth), Motrin (McNeil), and other trade names.

Trade names are used to uniquely brand an individual com-

Table 8.3 *Drug names*

Chemical / Biological	Scientific description of drug compound
Generic	Simplified name based on drug function
Trade name	Branded name, protected by trademark law

Table 8.4 *Selected generic drug naming conventions*

Name element	Category	Examples
-vir	antivirals	acyclovir, combivir
-mab	monoclonal antibodies	cetuximab (ImClone's Erbitux), rituximab (Biogen Idec's Rituxan)
-rsen	antisense oligonucleotides	fomivirsen (Isis Pharmaceutical's Vitravene)

pany's version of a drug. A drug's proprietary or trade name must be approved by the FDA and cannot imply efficacy. Trade names are protected by trademark law, preventing generic companies from using them even after patents expire, and encouraging pioneers to develop strong brand identity to extend their market dominance past patent expiration. Despite the restriction that trade names cannot imply efficacy, drug makers often select names with connotations aligned with a drug's intended use. Vick's Dayquil and Nyquil respectively suggest daytime or night time tranquility in treating cold and flu symptoms.

DEPARTMENT OF AGRICULTURE

The USDA regulates plant pests, plants, and veterinary biologics. Because of the potential for plants to interact unfavorably with the environment, it is important to determine that a plant is safe to grow outdoors in order to prevent the release of plants that may harm other plants and animals or spread like a weed.

Several USDA agencies are involved in regulating and monitoring the use of biotechnology for agriculture. The Agricultural Marketing Service is responsible for administering plant variety and seed laws. AMS also offers laboratory testing services for genetically engineered food and fiber products. The Agricultural Research Service (ARS) is USDA's in-house science agency. ARS works to improve the quality, safety, and competitiveness of U.S. agriculture. The agency's biotechnology research includes introducing new traits and improving existing traits in livestock, crops, and microorganisms; safeguarding the environment; and assessing and enhancing the safety of biotechnology products. ARS also develops and provides access to agricultural resources and genomic information.

The Food Safety Inspection Service has responsibility for the safe use of engineered domestic livestock, poultry, and products derived from them. The Animal and Plant Health Inspection Service (APHIS) is responsible for protecting agriculture from pests and diseases. APHIS regulates the movement, importation, and field testing of genetically engineered organisms through permits and notification procedures. In addition, APHIS Veterinary Biologics inspects veterinary biologic production establishments and licenses genetically engineered products. Another unit of APHIS, the Biotechnology Regulatory Services, monitors biotechnology industry trends and forecasts scientific advancements to better help regulate the industry.

The Foreign Agriculture Service supports the overseas acceptance of biotechnology and crops that have been reviewed by the U.S. government agencies to support U.S. farm exports and promote global food security. The Economic Research Service conducts research on the economic impact of genetically engineered organisms. The Cooperative State Research, Education, and Extension Service administers the biotechnology risk assessment program as well as research programs in gene mapping, sequencing, and biotechnology applications.

ENVIRONMENTAL PROTECTION AGENCY

The EPA regulates microbial and plant pesticides of chemical and biological origin, new uses of existing pesticides, and novel organisms that may have industrial uses in the environment.

Under the Federal Insecticide, Fungicide, and Rodenticide Act, EPA regulates the domestic manufacture, sale, and use of all pesticides, including those derived through modern biotechnology. Section 408 of the Federal Food, Drug, and Cosmetic Act gives the EPA the mandate to ensure that any pesticide residue in or on a food product falls within a safe limit. Either a safe limit (known as a food tolerance) is set, or an exemption from the requirement of a tolerance is granted. The Toxic Substances Control Act Biotechnology Program tracks all chemical substances or mixtures of chemical substances produced or imported into the United States. Genetically engineered microorganisms meet the program's legal definition of a "mixture" of chemical substances and are therefore regulated under this act.

IV

The Business of Biotechnology

This section describes the commercial considerations of biotechnology companies, with the goal of enabling their effective management. The long development timelines and large up front capital investments required by biotechnology firms create unique challenges. Companies need to raise large sums of money—in several rounds—long before they are able to generate positive revenue streams.

Biotechnology companies must find a balance of resource allocation that combines a strong scientific base, sufficient financing, and relevant business expertise.

Success in biotechnology absolutely requires three elements: strong management, financing, and technology. Management is arguably the most important ingredient because of the vital need for proactive marketing, to maintain financial health, and to redirect or replace research efforts when necessary. In the absence of good management, otherwise sufficient financing and technology cannot realize their potential.

Beyond simply obtaining financing and other resources, and managing expenditures, management is also responsible for aligning research efforts with market needs. A common reason for commercial failure is development of products, which are often based on innovative science, for which profit-enabling markets do not exist.

Because most biotechnology applications are based on innovation rather than price competition or modest improvements of existing applications, it is often necessary to invest heavily in research and development long before any revenues are generated. Intellectual property protection is therefore very important in biotechnology, securing the potential to achieve profitable revenues.

The knowledge-based and research-intensive nature of biotechnology companies gives them unique characteristics. The speed of technological development and the fundamental role of people and ideas make biotechnology companies harder to assess as investment opportunities. Startups require growth capital at an early stage because retained profits, if they exist, are usually insufficient to support the characteristically long development times. Additionally, as a company matures the management team will require different sets of skills to satisfy changing needs; early stage companies require multifaceted individuals to drive research and build relationships, whereas more mature companies need specialists who can manage relationships and support the development and commercialization of research leads.

Chapter 9
Biotechnology Company Fundamentals

I like to buy a company any fool can run because eventually one will.
Peter Lynch, Fidelity Magellan Fund Manager

The unique conditions faced by biotechnology companies—the necessity for funding well in advance of revenues, need for strong intellectual property protection, and regulatory clearance requirements—influence individual companies and inter-company industry relationships. This chapter profiles some common characteristics in the structure of biotechnology companies to establish a framework for the later discussion of operational elements.

COMPANY FORMATION

There are several ways that new biotechnology companies form. Two common themes are a scientist in an academic or industrial position forming a new company to commercialize a novel idea, or an established company or university forming a new company to develop a potentially lucrative technology. In other cases, a company may be formed by entrepreneurs starting companies to pursue lucrative opportunities such as developing validated drug leads, finding new uses for existing drugs, or exploiting new scientific advances.

Box

Speedel: Spinning off to develop a shelved drug

Speedel was formed in 1998, in the wake of the 1996 merger of Sandoz and Ciba–Geigy that formed Novartis, to focus on SPP100, a drug lead which was shelved by Novartis. Founder Alice Huxley, who had been working on the drug at Ciba-Geigy prior to the merger, convinced Novartis' management to spin-off Speedel.

In licensing the drug lead to Speedel, Novartis retained a buy-back option, which they exercised in 2002. SPP100 received FDA approval and was launched as Tekturna in 2007. Besides being Speedel's first drug, Tekturna is also the first in a new class of drugs for high blood pressure, blocking kidney renin production (which can raise blood pressure).

One of Novartis' primary motivations in abandoning SPP100 was concerns about the ability to manufacture the drug profitably. Speedel was able to overcome the high manufacturing costs by developing a new synthetic process. This arrangement shielded Novartis from the manufacturing process developmental risks and lilely granted Speedel scientists greater operational flexibility.

Interestingly, Speedel later launched a dispute over claims that Novartis failed to pay them royalties on use of their manufacturing technology. In July 2008 Novartis announced plans to acquire Speedel for an estimated $880 million.

START-UP

Independent start-ups emerge from the union of a scientist or group of scientists with a commercial vision, and one or more financiers. The prototypical start-up involves a group of scientists with proven expertise who seek to commercialize a new technology. While a small amount of initial funding may be provided by "friends and family," the bulk of early capital comes from professional investors such as venture capitalists and angel investors. Professional investors also contribute valuable business expertise, guidance and networking in addition to the money they provide in exchange for equity. Because their ultimate goal is to gain a return on their investment—a goal which may be at odds with the founder's primary interests—it is necessary to align the goals of investors and founders. This topic is discussed in further detail in Chapter 10.

Genentech, founded in 1976 by venture capitalist Robert Swanson and biochemist Herbert Boyer, is an example of a biotechnology company that was initiated as an independent startup. In 1973 Boyer and fellow biochemist Stanley Cohen demonstrated the first expression of spliced genes in bacteria. Swanson approached Boyer with the proposal to form Genentech to commercialize this revolutionary technology. After failing to secure funding from the National Institutes of Health, Genentech turned to venture capital firm Kleiner and Perkins.

Staging their investments over milestones, Genentech's first goal was to validate the potential of gene splicing by demonstrating the ability to produce a human protein in bacteria. Following proof-of-principle production of the brain hormone somatostatin, Genentech later produced recombinant human insulin, the first marketed biotechnology product (see Box *Genentech: Commercializing a new technology* in Chapter 2).

Genentech wasn't able to reap the full rewards of its first innovative product. In order to obtain the financing necessary to develop human insulin, Genentech had licensed manufacturing and distribution rights to Eli Lilly. Three years later, Genentech became the first biotechnology company to independently manufacture and market a product as they leveraged their proven developmental expertise and a better financial position to introduce Protropin (human growth hormone). Today Genentech is one of the largest biotechnology companies.

Technology Transfer

In academic settings technology transfer refers to the transfer of research findings from academic laboratories to the commercial marketplace for public benefit.

The rationale for technology transfer is that the various institutions involved in biotechnology product development are stratified in their specialties. Academic labs, for example, are well positioned for basic research. Academic researchers are rewarded primarily based on their research performance, as measured by publication of cutting-edge science, and not by the commercial value of their outputs.

This focus on basic research makes academic researchers poorly positioned for applied research. Whereas academic researchers may be primarily interested in elucidating the mechanism of a biological interaction, a commercial researcher may be more interested in the potential to modulate an observed interaction, or in the potential for lead compounds to pass safety and efficacy tests in pre-clinical and clinical trials.

Stratification also exists within commercial research. Smaller companies generally lack the resources for large-scale manufacturing or clinical trials, leading them to focus on late-stage basic research and early-stage applied research, and transferring strong leads to larger companies for further development.

BUSINESS MODEL

Biotechnology companies can pursue research in one or more of the general categories of medical applications, agricultural applications, and industrial applications (see Table 9.1).

Biotechnology firms are distinguished from pharmaceutical firms and firms casually using biotechnology techniques by their intense research focus and the emphasis on molecular biology techniques. While the prototypical biotechnology firm focuses on drug development, applications range in diversity from using bacteria to decompose oil spills to using genetically engineered bacteria to producing spider silk in goat milk (see Chapter 6).

Within the aforementioned application categories there are five basic activities in which biotechnology companies engage: basic research and target discovery; applied research and lead refinement; clinical and prototype research; manufacturing; and, sales and distribution (see Figure 9.1). Vertical integration was once a reality for pharmaceutical companies and a goal for emerging biotechnology companies, but a number of factors have made it more efficient for companies to specialize in just a few elements of the research-development-commercialization path. Most companies focus on just one or two of these activi-

Table 9.1 *Biotechnology application categories*

	Red: Medical biotechnology
Description	Drugs and other agents to treat, cure, or prevent disease, and products that assist in the diagnosis of diseases or measurement of critical factors in health and disease.
Characteristics	High up-front development costs, FDA (or other regulatory) approval required prior to sale. High post-approval profit margins.
	Green: Agricultural biotechnology
Description	Products and applications related to livestock and crop production, and agricultural production of biotechnology products.
Characteristics	Development costs are often similar to drugs, profits are often lower.
	White: Industrial biotechnology
Description	Modification or improvement of industrial processes or performing tasks previously served by industrial processes such as paper processing, bioremediation, or synthesis of organic compounds.
Characteristics	Reduced regulatory burden decreases development costs.

See chapter 6 for details on specific applications

ties, with a select few vertically integrated companies engaging in most or all of these activities. In the case of drug development, the costs associated with development and commercialization activities prohibit all but the largest biotechnology and pharmaceutical companies from integrating all the components. Even large vertically integrated companies outsource selected operations to smaller specialized firms.

Further dividing the above biotechnology industry segments, biotechnology companies can be segmented based on whether they focus on selling products or services. Nearly all biotechnology companies sell products and services to other companies, as opposed to selling directly to consumers. Products include physical items such as drugs, tools, reagents and other manufactured compounds. Service firms perform de-

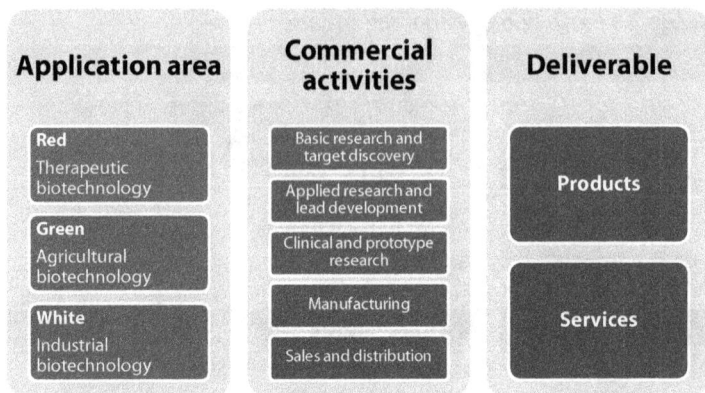

Application area	Commercial activities	Deliverable
Red Therapeutic biotechnology	Basic research and target discovery	**Products**
	Applied research and lead development	
Green Agricultural biotechnology	Clinical and prototype research	
White Industrial biotechnology	Manufacturing	**Services**
	Sales and distribution	

Figure 9.1 *Biotechnology company activities*

fined services rather than selling defined products. Examples of services include research support activities such as contract research, manufacturing, lead-optimization, and diagnostic services such as paternity and forensic testing.

Some companies combine product development and service offerings in a hybrid model, selling proprietary services while using them internally for product development as well. This strategy can potentially demonstrate the value of a company's scientific foundations and defray the high costs of product development, but it is also subject to abuse. Companies with faltering core elements may adopt a hybrid strategy to leverage their intellectual property and potentially distract investors; platform and service firms can add a drug development unit to distract investors from technologies facing obsolescence, and product-development companies with failed leads can re-deploy their resources by selling services and technologies based on their proprietary techniques and knowledge. A crucial question to ask of hybrid firms is whether their business model makes sense as part of a long-term strategy, or is potentially being implemented primarily to buy time or attract additional investors.

COMPANY CHARACTERISTICS

The objectives of biotechnology companies and the challenges they face are reflected in their design and business structure. Biotechnology business development requires three elements: research proficiency to enable development, funding to support development, and a competitive advantage to enable profitable commercialization. It is the role of management to bring together these necessary elements and guide commercial activities to a productive end. A company's location, size, and maturity are other important characteristics that influence business operations.

LOCATION

In selecting a business location, biotechnology companies must consider factors such as access to capital and skilled workers, costs of doing business, regional laws, and regulations. The accumulation of biotechnology in an area enhances access to skilled workers and capital and leads to the development of biotechnology clusters.

The influencing factors in biotechnology company site selection change as a company matures. Start-up companies focus on research and development and are therefore more motivated by factors enabling research, such as access to skilled workers and funding, than by tax credits or the cost of research space. Mature companies with sustainable operations are able to attract capital and employees from distant locations and accordingly place more emphasis on factors affecting their bottom line, such as operational costs.

Zoning is also a significant issue. Biotechnology companies may work with dangerous chemicals, radioactive materials, or perform animal testing. The inability to receive permits for these critical elements can be a non-starter in location decisions.

One of the most important needs of young biotechnology companies is access to skilled workers, who tend to aggregate in areas with high concentrations of biotechnology companies or academic research laboratories. Because biotechnology research is risky by nature, employees must consider their options in the event that their job is terminated. Employees of biotechnology companies often have specialized expertise, which makes the difficulty of finding a new job in the event of downsizing or restructuring an important consideration. Having several potential employers in a region, such as biotechnology companies, service providers, and universities, can ease transitions for employees. Biotechnology companies without nearby neighbors may therefore find it difficult to recruit top-flight researchers.

The need to offer workers career security causes biotechnology start-ups to congregate in close proximity, often near sources of skilled workers. Because the best places to find top workers are areas with high quality universities, medical centers, and research centers, it is not surprising that many biotechnology companies have formed in the vicinity of high concentrations of academic institutions and medical and research centers.

MATURITY

In biotechnology, stability comes with size and maturity. Biotechnology companies can be roughly divided into three groups according to these qualities. While the distinction between these groups is somewhat subjective, it provides a useful framework to distinguish between the various stages in the maturation of a biotechnology company.

Mature companies are established large-cap biotechnology companies and have relatively stable revenue streams; promising companies have strong fundamentals and excellent prospects for near-term profitability; emergent companies are nascent biotechnology companies without a clear path to profitability, making it difficult to reliably assess their long-term potential.

Mature companies are profitable. Compared with promising and emergent companies, the revenues that these companies earn permit them to invest relatively more resources in research and development programs. While they still face the inherent risks of biotechnology product and service development, mature companies will fare better than smaller ones in the event of

Table 9.2 *Assessing biotechnology company maturity*

	Profits	Products
Mature	Measurable and significant	Products on market and in development
Promising	Small profits or strong prospects for near-term profitability	Products on market, emphasis on products in late-stage development
Emergent	No clear path to profitability	Developmental uncertainty

a research disappointment or in unfavorable market conditions. These relatively stable companies lend themselves to traditional financial analysis far more readily than promising or emergent biotechnology companies.

Promising companies either have products and services on the market with growing sales, or have potential big sellers in late-stage development or clinical trials. A well-positioned promising company may be unprofitable, but it will have sufficient cash reserves to fund research and operations until profits emerge. Some of the companies in this group will go on to become industry leaders while others may fail, remain small, or be acquired by competitors. Promising companies involve significant risk.

Emergent companies are still seeking to commercialize their first product or service or are struggling to gain acceptance for their technology platforms. These companies may announce that they have products in development which address lucrative market opportunities, but on closer examination it often becomes clear that these products are years away from completing development. Aside from the risk of developmental failure, a second risk is that competitors may emerge or market trends may shift and these emerging companies will be in the uncomfortable position of having devoted their resources to developing a product for which the market potential has diminished. Eventual success for these companies is far less certain than for promising companies.

Chapter 10
Finance

> Where else can you combine making money—which is
> the primary purpose of venture capital—with the feeling
> that you're doing some good in the world?
> *Venture capitalist Frederick Adler, regarding biotechnology*

B iotechnology companies employ a variety of methods to
fund the expensive research and development programs
that are necessary to produce marketable products and
services. Sophisticated machines and well-trained workforces
cost a lot of money. The requirement for precision and consistency demands refined tools and reagents of great purity which
are expensive to purchase, maintain, and operate. The cost of
attracting and retaining the highly skilled workers necessary
for cutting-edge research is also significant. The research and
development required to produce biotechnology products and
services takes a long time and is fraught with unexpected setbacks, requiring substantial quantities of funding long before
revenues can be anticipated.

These high capital requirements, combined with uncertain-
ties about market size and projected revenues, and the relative
lack of tangible assets to use as collateral, prevent the extensive
use of conventional debt financing, particularly for biotechnology companies that do not have a proven product. *In lieu* of
debt, biotechnology companies often exchange equity with professional financiers or sacrifice some of their autonomy to sponsoring partners in order to finance research and development.

Funding for biotechnology companies comes from various

sources. Professional financiers exchange money for equity in promising companies with the hope of realizing a return on their investment. Established companies can spin-off new entities to exploit promising research projects, contract elements of research to firms with specialized expertise, or form partnerships with innovative companies and provide cash in exchange for research leads or the right to market developed products. Various federal, state, and regional programs also provide funding opportunities to biotechnology companies. Industry estimates of the sources of capital for the first decade of a biotechnology company's existence are that 10 percent comes from venture capital and other private equity sources, 40 percent from public markets, and 50 percent from senior partners.[1]

DEVELOPMENT STAGES AND FUNDING

A developing biotechnology company will go through several discrete stages during its growth. Regardless of a company's origins—whether it is spun-off from a larger firm or formed to capitalize on technology transfer from academic labs (see *Technology Transfer* in Chapter 9)—the same stages exist. In some cases, early funding requirements may be met through the sponsoring institution prior to the actual founding of the nascent company.

The general framework described below considers a product development company, although the discussion is relevant to service companies as well. While this presentation makes the process appear linear, it seldom is. Biotechnology companies need to continually investigate lucrative tangents emerging from research, to simultaneously pursue multiple opportunities, and to adapt business and research plans to accommodate changing internal and external influences.

Funding is commonly tied to discrete milestones, which should provide objective measures of progress toward commercialization and profitability. The completion of milestones pro-

---

1   Hess, J., Evangelista, E. Pharma-biotech alliances. *Contract Pharma*, September 2003.

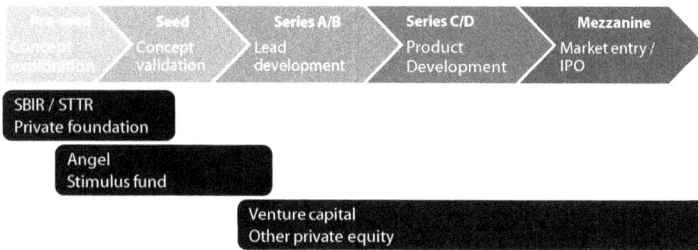

Figure 10.1 *Funding stages and sources*

vides evidence that the goals are attainable, reducing risk and thereby increasing the valuation.

In the seed stage, a biotechnology company simply has an idea for a product. There is an immediate need for sufficient funds to develop a proof-of-principle so that the company can attract further funds and develop a prototype. The funds that launch a company, fund its proof-of-principle research, and generally support an early assessment of business feasibility, are called seed funds. This initial stage is generally the most dilutive, with financiers commonly seeking ownership of 50 percent or more of the company (see Table 10.1).

While venture capitalists once commonly provided seed investments, seed stage funding is now more commonly provided by angel investors, friends and family, and bank loans (which may be personally guaranteed by the founders). As a company develops, each financing stage is generally less dilutive than the previous stages. Within a short time after seed funding, generally six months, a company should be able to demonstrate proof-of-principle. This often coincides with patent filing, which is the first step to securing market protection for the product in development.

The expansion stage involves production of prototypes and requires more money than the proof-of-principle stage, granting a relatively smaller share of equity for the amount of money invested (see Table 10.1). This stage may see several rounds of funding, often named by how many external fundings have occurred. The seed round is named first-stage. Series A, series B, and subsequent rounds are named accordingly. These post-seed funding rounds may each raise millions of dollars and should support company operations until commercialization occurs.

The final private funding stage, mezzanine funding, is used to prepare products for market and enable a company to independently survive on revenues and loans, or to launch an initial public offering (IPO). As an alternative to public offerings, companies may be acquired by other private or public firms.

PRIVATE EQUITY

The types of individuals and groups engaged in financing a biotechnology company change as the company grows. These investors vary in their tolerance for risk, their expectations for returns, and their financial resources. Figure 10.1 shows the general alignment of funding stages with funding sources. Private equity investors, described in this section, make investments in promising private firms in exchange for a share of equity, or ownership.

FRIENDS AND FAMILY

While seed stage funding may be obtained by offering equity to friends and family or seeking loans, later stages require professional financiers who are able to raise larger funds, assess risk, and provide guidance and other services. These professional investors include angel investors, venture capitalists, merchant banks, investment funds, incubators, and corporate partners who finance and share business expertise and contacts with entrepreneurs.

With previous experience in developing or funding biotechnology companies, financiers can help growing companies avoid pitfalls. Moreover, unlike other sources of guidance such as consultants, professional financiers have a long-term vested interest in a company's success. This long-term interest is exemplified by their compensation. Rather than charging for their services, the guidance and assistance of professional financiers are sometimes viewed as justification for the opportunity to

Table 10.1 *Funding stages and investment sizes*

Funding stage	Investment	Percent ownership
Seed	$300-600k	40-60
First-round	$1-5 million	40-60
Early mezzanine	$5-15 million	20-30
Late mezzanine	$20-50 million	25-35

Source: Birndorf, H.C. Rational financing. *Nature Biotechnology*, 1999. 17:BE33-BE34

invest in a company. The prestige and networking contacts associated with partnership with esteemed financiers can also facilitate additional funding, help secure top researchers and managers, and attract representation by leading investment banking and law firms.

ANGEL INVESTORS

Angel investors are wealthy individuals who invest in private companies. According to Securities and Exchange Commission guidelines, angel investors must be accredited investors; they must have an individual net worth exceeding $1 million or earn more than $200,000 annually (or $300,000 in combination with their spouse) to be accredited. Angel investors tend to have succeeded in their field and are looking for opportunities to help young entrepreneurs develop new companies. A significant difference between angel investors and venture capitalists is the scale and intensity of their activities. Unlike venture capitalists who manage venture partnership-raised funds, angel investors invest their own money, and traditionally make investments with less-onerous terms.

Because helping companies grow is usually not their full time occupation, angel investors often invest in fewer companies than venture capitalists and may spend less time actively involved in company development.

VENTURE CAPITALISTS

Venture capitalists are professionals who manage venture capital funds. Their goal is to invest money in promising companies in return for equity. Experienced in developing companies, venture capitalists offer their business expertise in exchange for the opportunity to invest in a company. In a competitive environment, venture capitalists may compete for the opportunity to invest in individual companies.

The investors in venture capital funds include wealthy individuals and corporate entities such as pension funds. In order to produce the return on investment that their investors seek, venture capitalists make a number of risky investments in the hopes that at least some of them will do phenomenally well. To produce high returns, venture capitalists must take a large amount of equity and frequently demand board representation in a developing company in exchange for their investment.

EXIT

Entrepreneurs and financiers may cooperate in the development of a venture, but they have dissimilar goals. While entrepreneurs may desire to develop or commercialize a product or service, financiers are generally primarily motivated to gain a return on their investment. This does not imply that descriptions of financiers as money-hungry are always justified, but financiers must be permitted to profit from investing in a company. Investors seek to purchase equity in promising companies that reflects the risk of failure. They accept the risk that they will not recoup their full investment in exchange for the possibility that they will turn an appropriately sized profit.

Even if entrepreneurs and financiers are both primarily interested in financial returns, financiers usually have a liquidation preference, enabling them to obtain the first proceeds from any sale of the company. This liquidation preference is necessary to marry interests with the entrepreneur who, having invested less capital than the financiers, could otherwise simply sell their shares at a profit without building any value for the financiers.

Two common exits for investors are initial public offerings (IPOs; described later in this chapter) and sale of private companies through merger or acquisition. IPOs are often favored by founders, as they enable a company to maintain its independence, but acquisitions may offer greater returns. An analysis of biotechnology transactions from 2003-2005 found that median return for IPOs was 2x, while acquisitions delivered a median return of 3.5x.[2] While the choice to pursue a merger/acquisition or an IPO is complex and influenced by current market dynamics, a company's business model, and other elements, this observation does illustrate that IPOs are not necessarily the preferred option.

PUBLIC MARKETS

Public listing is an important factor from an investor perspective because it permits the realization of gains on investments through the sale of shares granted for investments. Most investors are unwilling to wait for residual revenues from cashflow, and seek liquidity opportunities such as selling their equity shares in public markets or to an acquiring firm.

Companies sell shares in public markets to access greater sums of money than would be available through private markets, as well as far greater dissemination of shares among investors, which typically leads to fewer restrictions on the company, without a control block or lead investor. In raising money, private markets operate very much like public markets; investors provide cash in exchange for shares. Fluctuations in private and public market conditions can affect the number of shares that investors expect to receive for a given investment. A fundamental difference between public and private markets is that public markets are more liquid—it is easier to sell publicly-traded shares than privately-issued shares (see the requirements to be an accredited investor in private equity in the discussion of *Angel Investors* above). The relative inability for investors to divest

2   Behnke, N., Hültenschmidt, N. New path to profits in biotech: Taking the acquisition exit. *Journal of Commercial Biotechnology*, 2006. 13:78-85.

themselves of investments in private firms leads them to expect a price discount. Consequently, public firms tend to carry higher valuations than similarly positioned private firms. The ease of investment in public companies also enables investors to readily invest in undervalued firms, permitting rapid recovery of share prices.

Public registration also offers many advantages for companies. The greatly increased liquidity (and objective value) of publicly traded shares relative to privately traded shares facilitates their use *in lieu* of cash. Shares, rather than cash, can be used to form partnerships or to raise equity, and possibly acquire, other companies. Additionally, granting time-restricted stock purchase options to employees and constituents can also reduce cash expenditures and improve retention. Public registration can also grant credibility with investors, since public companies have increased reporting requirements and must file audited financial statements, and can provide an objective measure of a company's valuation. Because investors in private companies often anticipate public offerings as a means to realize gains on their investments, the potential of a public offering can attract funding while a company is private.

INITIAL PUBLIC OFFERING

An initial public offering (IPO) is the first sale of stock by a private company to the public, marking the transition from being privately traded to publicly traded. Unlike the exchange of shares in the open market following the IPO, cash proceeds from the initial offering (less underwriter discounts and expenses) goes to the listing company, providing an opportunity to raise significant amounts of capital.

An IPO requires interest on the part of the public to purchase stock in the newly listed firm. The concept of IPO windows is central in engaging in an IPO. The window is considered open in a given sector when markets are receptive to IPOs. Failure to obtain sufficient interest in the "roadshow" or marketing of the IPO by the company and its underwriters may result in cancellation, or postponement with a potentially decreased offering price and size. Preparation for an IPO can be lengthy and expensive, and may distract management from other responsibilities, so careful timing is required to ensure that preparatory activities can be performed in advance of window opening, and before it closes.

OTHER FUNDING SOURCES

Private companies are restricted in the ways in which they can issue shares, and the lack of liquidity of private shares devalues them in comparison to publicly-traded shares. Many private companies are also too small, or have too much uncertainty in their future prospects, to attract equity investments. These companies must look to alternate sources of funding.

GOVERNMENT FUNDING

In addition to the numerous sources of private funding, there are several government programs that can support biotechnology projects. Funding that supports the early stages of research that may lead to drug development is usually obtained from public granting institutions such as the National Institutes of Health, National Science Foundation, and others. Additionally, regional interests such as local and state governments and other development initiatives and private institutions can offer land grants, temporary tax and public utility waivers, loans, and cash. An advantage of government funding relative to private financing is that equity is not exchanged for funding, leaving ownership equity undiluted. The review process to receive government funding can also act as a proxy for private investors, assuring them that objective outsiders have faith in the scientific and commercial prospects of a company.

CORPORATE PARTNERSHIP

A significant source of revenue for biotechnology firms is other firms. Small firms may form alliances and partnerships with more established firms, giving them access to development, regulatory, and sales expertise that can help bring a product to market. Senior partners may offer capital and business services in exchange for product marketing rights, patent licenses, or equity in a smaller firm.

The nature of corporate partnerships can be materially different from simple equity funding. While seasoned investors can contribute experience and networks to a growing firm, their direct involvement is often limited. Corporate partners may bring actual operational assistance. Few investors have much experience with operations such as clinical trials, but corporate partners may have extensive experience in designing and managing trials, and in submitting applications to the FDA. These direct benefits may be far more valuable than cash investments.

FOUNDATION SUPPORT

Many diseases have foundations dedicated to raising funds for their treatment and cure. International development foundations also sponsor research into treatments for diseases endemic to developing countries. Other foundations, like the Bill & Melinda Gates Foundation and the Archon X PRIZE for Genomics (related to the Ansari X PRIZE that awarded the first private development and launch of a spacecraft) focus on specific technical challenges. These funds can be tapped by academic researchers or growing biotechnology firms to advance research. In addition to financial support, foundations can also provide valuable networking, advice, prestige, and publicity.

Chapter 11
Research and Development

*I am not discouraged, because every wrong attempt
discarded is another step forward.*
Thomas Edison

Because biotechnology is focused on commercializing innovation, research and development are central to any company's strategy. R&D is central in biotechnology because it is the means by which companies develop innovative products which are worthy of patent protection and which can sell for premium prices.While some companies may eschew research and focus instead on developing or commercializing leads produced by others, R&D is still critical to their success—even if they are not directly involved. This chapter focuses on the process and strategies of R&D.

As described in the chapters on marketing, R&D is not the endpoint for biotechnology companies; developed products must address meaningful needs of customers who are able to pay profit-enabling prices.

NON-DRUG BIOTECHNOLOGY

An important distinction between drug development and other biotechnology applications is that drugs are highly regulated—during development, and after they are in the market. While this high degree of regulation can be a burden to drug developers, it can also provide some substantial benefits. The third-party measures of clinical progress and therapeutic indications for which a drug is effective, provided by regulatory authorities, can be leveraged to facilitate financing or sale of drug leads.

R&D STAGES

Drug discovery is described in greater detail in Chapter 4, and the general scheme of R&D is summarized here. There are two fundamentally different kinds of research: basic and applied research. Basic research is directed at improving fundamental knowledge, whereas applied research is directed at applying knowledge gained from basic research. Different toolsets, mindsets, and players are involved in these types of research. Basic research does not directly produce biotechnology products. Instead, it lays the foundations upon which products are developed. Applied research is required to enable further development.

The basic stages for drug and non-drug development are shown in Figure 11.1. The stages are segregated by fundamental differences in operations and objectives. For example, different mindsets, skill sets, and deliverables are involved when discovering a potential drug versus refining the properties of a drug lead.

While the stages are presented here as a discrete linear process, in practice the distinctions between stages are not so distinct. Promising leads may be advanced to development while discovery-stage research continues to investigate basic-science questions and additional opportunities. Likewise, commercialization activities need to be initiated while discovery-stage research is occurring. Waiting for FDA approval before establishing manufacturing capacity or initiating reimbursement discussions with healthcare payers is not practical due to the amount of time required to prepare for these activities.

DISCOVERY

Early stage R&D focuses largely on defining the system being examined and identifying lead compounds. At this stage the primary objective may be to simply study a disease mechanism

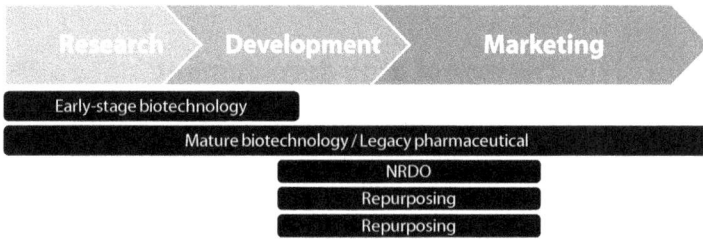

Figure 11.1 *R&D Stages*

in the search for druggable opportunities, or to identify compounds that interact with disease mechanisms. The high rate of attrition of leads emerging from discovery-stage research is a motivation for many companies to license leads from others, or to form research alliances. *Outsourcing Innovation*, later in this chapter, describes some other methods used by established firms to expand their access to novel research.

DEVELOPMENT

Once leads have been identified and have passed early tests to demonstrate promise, it is necessary to ask a new set of questions. Development-stage researchers strive to refine lead compounds, seeking to expand the understanding of how they work, to investigate opportunities to improve stability and efficacy, and to reduce potential side effects. The development stage is also where pre-clinical and clinical trials occur (see *The Five Basic Steps of Drug Development* in Chapter 4) to determine the safety and efficacy of the drug in humans.

During the development process the market potential of a biotechnology product also needs to be assessed. The target markets need to be identified, and the suitability of a product to serve these markets must be evaluated. The quality of a market itself must also be assessed, and potential reimbursement issues should be identified early. Marketing and reimbursement are discussed in greater detail in Chapter 12.

The cost of drug development

When assessing the cost of drug development, most sources cite one of two studies from the Tufts Center for the Study of Drug Development (CSDD). In 2003 the CSDD estimated the average cost of developing an approved traditional pharmaceutical drug to be $802 million, and measured the development time to be 10-15 years.[1] In 2007 they estimated the cost of biologic drug development to be $1.2 billion, with a development time greater than 12 years.[2] It is worth noting that the biologic cost estimate derives from a relatively small sample size: 17 investigational drugs from four firms. These estimates are valuable because they measure the overall cost of drug development programs, but it is important not to use them as metrics for the cost of developing individual drugs.

These estimates do not suggest that a company would need to invest $1.2 billion, or even $800 million to develop a new drug. First, the estimates include the opportunity-cost of capital, which is an expression of the lost income opportunities because investments must be made years in advance of approval—the opportunity cost of not being able to invest money in stocks, bonds, etc. Actual cash outlays for biologic and small molecule drugs were respectively estimated to be $500 million and $400 million. Accordingly, it is not accurate to state that biologics cost $1.2 billion to develop over 12 years—the duration of drug development is already expressed in the cost estimate. Second, these estimates include development costs for failed drugs. They do not reflect the cost to develop a new drug; they measure the cost of drug development programs. The estimates were derived by assessing the cost of drug development programs and dividing that cost by the number of products ultimately approved.

Therefore, while the average time-adjusted cost of biologic development drug has been estimated to be $1.2 billion, the cash expenditures required to develop a single new biologic drug are less than $500 million.

1 DiMasi, J.A., Hansen, R.W., Grabowski, H.G. The price of innovation: New estimates of drug development costs. *Journal of Health Economics*, 2003. 22:151–185.

2 DiMasi, J.A., Grabowski, H.G. The cost of pharmaceutical R&D: Is biotech different? *Managerial and Decision Economics*, 2007. 28:469-479.

Expected Cost

Duration

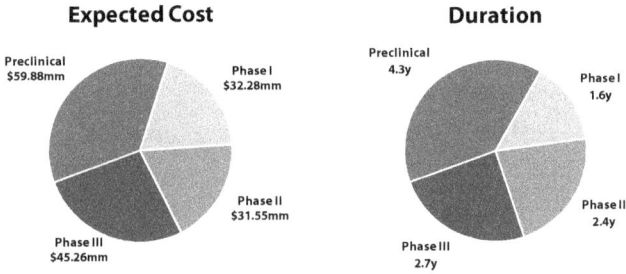

Figure 11.2 *Estimated cost and duration of biotechnology drug development*
Source: DiMasi, J.A., Grabowski, H.G. The cost of pharmaceutical R&D: Is biotech different? *Managerial and Decision Economics*, 2007. 28:469-479.

COMMERCIALIZATION

Once a drug has received FDA approval (or, in the case of other biotechnology applications, the product has completed development) it is time to scale-up production and start selling and distributing the product. As mentioned above, preparations for commercialization need to be initiated well in advance. Limited patent life and competition encourage companies to move quickly once products are ready for sale.

Commercialization activities also extend beyond simple manufacturing, marketing, and selling of products. Once a product with established safety and efficacy is on the market, it is time to start leveraging those attributes. Repurposing, described under *R&D Strategies* below, seeks to find novel applications of established drugs. A drug that is safe for one indication is likely to be safe for other indications: the challenge is to find additional applications where it is effective. Reformulation strategies, also described below, can extend a product's life after patents have expired by enabling the filing of new patents on improvements in delivery systems or efficacy.

R&D STRATEGIES

The essence of research and development is that it is the process by which commercial ideas are "de-risked" through testing hypothesis, and developed into commercial products. Drugs are the prototypical application of biotechnology companies, but R&D is also necessary for applications in other areas such as research tools, agricultural, or industrial biotechnology. Selected R&D strategies are discussed in this section. A general overview of biotechnology business models is presented in Chapter 9.

REPURPOSING

Repurposing is the search for new purposes for existing drugs, generally performed by a company other than the originator of the drug. The same basic metabolic processes may be implicated in multiple diseases, and a single drug may also have more than one biological effect, enabling one drug to treat numerous conditions. Seeking approval for additional indications can also potentially reduce clinical trial burdens by citing known safety profiles, and may also leverage existing patent protection. Advanced biotechnology techniques like expression profiling using microarrays can help screen for additional applications of existing drugs. High-profile repurposed drugs include thalidomide (see Box *Repurposing thalidomide*), Viagra, Minoxidil, and Propecia.

COMBINATION AND REFORMULATION

Combination and reformulation strategies can improve the therapeutic properties of drugs. Because novel combinations and reformulations may be patentable, these strategies are also vital tools in extending patent-protected life. Because these strategies overlap—some reformulations are combination products—they are described together.

Box

Repurposing thalidomide

Thalidomide was originally developed as a treatment for morning sickness in pregnant mothers. The drug was never approved in the United States, but in the late 1950s and early 1960s it was sold in 46 countries as a sleep aid and treatment for morning sickness in pregnant women. Thalidomide was recalled following the discovery that it caused severe birth defects. Although it is a small-molecule drug, not a biologic, thalidomide provides a case study for the potential of repurposing.

The cause of thalidomide-related birth defects lies in the chemical structure of the molecule. Thalidomide exists in two different chemical forms which are mirror-images of each other, (R)-thalidomide and (S)-thalidomide. They have identical chemical compositions but different shapes, and readily convert between forms. The (R) variant is effective against morning sickness, but the (S) form causes birth defects.

In further studies on the biological activities of thalidomide, it was discovered that the drug had applications beyond morning sickness. Based on the observation that U.S. AIDS patients were illegally importing thalidomide to treat wasting, Celgene, which started as a spin-off from a chemical company, sought to seek approval of the drug for this indication. The rationale was that despite the drug's tragic history, the FDA would likely approve the drug for AIDS patients.

Facing setbacks in demonstrating efficacy for HIV treatment, Celgene received FDA approval in 1998 to market thalidomide for leprosy under the brand name Thalomid, and later gained approval for multiple myeloma. The FDA has applied special restrictions for physicians, pharmacists, and patients (who must comply with contraceptive measures) to control off-label use and prevent birth defects. Thalidomid sales, which totalled $447 million in 2007, helped Celgene reach profitability in 2003, 17 years after its formation.

COMBINATION

As products mature, the motivation to combine them increases. After a drug has been on market for several years, the developer will have extensive knowledge on safety and efficacy well beyond that collected in initial clinical trials. Developing combination products provides an opportunity for biotechnology companies to leverage this knowledge and potentially extend patent life by developing new and useful drug combinations. Combination drugs have existed in a relatively simple form for decades. Combinations of antibiotics and HIV cocktails have provided multifaceted approaches to disease treatment. These work by targeting numerous disease processes simultaneously, and are generally intended to prevent infectious disease. Advanced combination therapies integrate elements with very different modes of action. Drug coated stents, for example, combine stents which physically prevent artery narrowing with drugs that prevent reblockage.

REFORMULATION

Altering a drug's formulation or delivery method can yield improvements in safety, efficacy, and ultimately result in preferential prescription and increased patient compliance. Just as repurposing can leverage existing safety data, reformulation can also significantly reduce the cost of developing improved drugs.

A pioneer in reformulation is Enzon, the fifth company to gain approval for a biotechnology-based drug. Enzon's Adagen, which gained approval in 1990, is a reformulation of adenosine deaminase enzyme (ADA), a drug for severe combined immunodeficiency disorder, with polyethylene glycol (PEG). PEG, which is used as a thickener and foam stabilizer in food and cosmetic products, can decrease immune-system reactivity and increase the blood circulatory time of drugs. Without PEG, ADA is ineffective.

A significant benefit of reformulation is that alterations may be patentable, enabling a company to secure rights to an improved version of a drug which lacks other patent protection. This strategy is commonly used to extend the lifespan of drug brands after the initial set of patents expire. The scope of patent protection in such a case is limited to the advanced formulation. Competitors are free to sell generic versions of the drug, but the holder of the formulation patent will be able to leverage their protected formulation as a marketing differentiator.

In an example of brand extension by reformulation, Pfizer was able to generate $8 billion in additional sales from 1990 to 1998 by developing Procardia XL, an extended-release hypertension drug. To develop the drug, Pfizer used Alza's OROS technology, which enables once daily dosing. Beyond simply reducing the number of daily dosages, the controlled release formulation also makes the drug more tolerable than immediate-release formulations, providing a significant marketable advantage.

In another example of reformulation, Abraxis Oncology improved on paclitaxel (the generic version of Taxol) by binding the drug to nano-scale protein particles. This reformulation provides significant patient benefits by eliminating the need for steroids or antihistamines, which are necessary to prevent hypersensitivity reactions to solvents required for other formulations.

SPECIALTY PHARMACEUTICAL / NRDO

The specialty pharmaceutical model historically referred to companies that acquired rights to drugs abandoned by large companies, usually due to insufficient revenue potential, and which then sought to gain approval for these drugs. The smaller size of the acquiring firms enabled them to profitably sell these drugs. Another variation was performing clinical trials to gain domestic approval for drugs approved in foreign markets. Specialty pharmaceutical also describes drugs targeted at specialist physicians or those requiring special handling or special disease-monitoring. Since the emergence of the "no research, development only" (NRDO) model in which companies license drug leads instead of developing them with internal research efforts, the two terms have merged. To avoid confusion, NRDO is used here.

The NRDO approach utilizes a quasi-virtual structure. NRDO companies have no internal research units and may also outsource later-stage activities. Two challenges shared by virtual companies and NRDOs alike are the inability to capture tangential discoveries that emerge in the course of research, and the need for an especially strong management team.

Many important discoveries emerge in the course of unrelated research. Because research is performed by external parties, NRDO and virtual companies may not be privy to subtle observations of potentially lucrative tangents. This loss of tangential opportunities is offset by a reduced risk profile. In purchasing drug leads that meet specific development require-

---

Box
Non-profit drug development

Drug development is synonymous with two financial outcomes: high cost and high revenue. These attributes stand in opposition to a non-profit business model. The great cost of drug development forces developers to focus on drugs that are likely to support sales which greatly exceed their development cost. This economic reality impedes the development of many treatments for acute conditions, conditions affecting small markets, and tropical infectious diseases that fall below the threshold of interest of drug developers.

Many promising drug leads with applications for minimally-profitable yet pressing conditions are shelved in mid-development by drug developers unable to justify the expense of completing development. By obtaining royalty-free licenses to develop these drug leads for neglected conditions, non-profit firms are able to use a NRDO model to meet these significant needs while for-profits can benefit from goodwill and tax benefits for their donations.

Leveraging the economies of scale and resources that have emerged from decades of aggressively funded drug development efforts, non-profit firms may raise money from individuals, corporations, and foundations, and utilize volunteer assistance from pharmaceutical scientists and other skilled professionals. One such case is the development of paromomycin, a treatment for visceral leishmaniasis, the second-most prevalent lethal parasitic disease after malaria.

The World Health Organization obtained rights to the injectable form of paromomycin, which had been shelved in mid-stage clinical trials, from Pharmacia (now Pfizer). Non-profit OneWorld Health then partnered with the WHO to shepherd the drug through clinical trials, and in 2005 received orphan drug approval for paromomycin from the FDA and EMEA.

ments, NRDO firms are able to shield themselves from much of the early-stage developmental risk inherent in drug development, although a premium price is often attached to this benefit. The lack of an internal research unit to tap for expert guidance places an additional burden on management to determine development agendas and assess opportunities.

OUTSOURCING INNOVATION

Large companies tend not to be very effective at innovation. Their structure, legacy, and incentive systems tend to favor a more conservative approach to business:

- There are more small companies than large companies. Even though small companies have fewer resources than large ones, by simple probability one would expect small companies to be responsible for a good share of innovation.
- Small companies can afford to take greater risks than large companies. A company with no, or meager, profits has only one option: to grow. Therefore, they tend to take greater risks than large firms, which sometimes leads to greater innovative output.
- Just as small companies may face bankruptcy if they fail to develop innovative products, managers at these firms are also less likely to advance in their careers unless their company is successful. A company-wide necessity for growth often results in greater tolerance for risk takers. By contrast, managers in large companies who back risky projects which fail may face censure or termination.

One means by which large companies can increase their output of new products is by licensing product leads and engaging in development partnerships with smaller firms. Even in cases where these strategies are more expensive than in-house innovation, the significant downside in terms of stock performance or individual careers often favors transferring the risk to outside parties—effectively paying money to reduce risk.

Beyond licensing and alliances, companies can also employ directed strategies to outsource innovation. A primary driver for outsourcing innovation is that many established companies have excess capacity to develop innovations, but cannot dedicate sufficient resources to perform early-stage research on all the possible opportunities. By outsourcing innovation to multiple partners who specialize in early-stage research, established firms can effectively reap the benefits of broad research programs without the associated loss of focus.

Chapter 12
Marketing

People don't buy products or services. They buy solutions to painful problems. If your customer has a headache, sell aspirin, not vitamins.
John N. Doggett, McCombs School of Business

M arketing is central to commercializing biotechnology. Although it is often thought of as a late-stage activity closely tied with advertising and sales, marketing plays an important role early in R&D. Marketing is one of the elements which distinguish biotechnology companies from traditional research labs. Biotechnology companies are not compensated by receiving grants to further productive research efforts; they are compensated by selling useful products to customers willing and able to pay profit-enabling prices. Accordingly, marketing plays a role in diverse activities such as guiding R&D and promoting, pricing, and distributing products.

MARKETING AS A GUIDE FOR R&D

R&D is unpredictable and, while R&D projects may have specific aims, it is not possible to actively control the outcome of R&D efforts. Unlike engineering projects, where a set of blueprints and plans dictate what the final product will be, it is not possible to choose what R&D will discover, and not all R&D products are worth commercializing. The challenges of guiding and predicting the outcomes of R&D create a pressing need to integrate marketing with R&D. Promising leads emerging from R&D must be evaluated for their market potential to ensure proper allocation of resources to promising projects and to avoid wasting time and money on less lucrative opportunities. The case of Pfizer's blockbuster Viagra is an excellent case for the value of integrating marketing with R&D. In the course of clinical trials for Viagra's original intended market, angina, a side effect of increased erections was discovered. Recognizing a potential to treat erectile dysfunction (a term popularized by Pfizer to replace *impotence* and facilitate marketing the drug), and aware that this market was poorly served, Pfizer proceeded with a small clinical trial with patients suffering from erectile dysfunction and found the drug to be highly effective. Buoyed by an aggressive marketing campaign, Viagra was prescribed more than 4 million times in the first week following FDA approval, and the drug quickly grew beyond $1 billion in revenues.

Beyond guiding R&D, marketing is also necessary to realize the full commercial potential of biotechnology. The notion that novel and useful products and services will attract customers in the absence of marketing is as untrue for biotechnology as it is for any other industry. It is imperative to conduct a thorough market assessment and develop a marketing plan to effectively profit from developing innovative biotechnology products. The market must be characterized in order to establish plans for, and effectively manage, the development and delivery of products and services that satisfy customer needs profitably. The case of Immunex's acquisition resulting from underestimating the market size of its first-in-class drug Enbrel is presented in a box later in this chapter.

Market evaluation is also important to fund and justify research efforts. It is the commercial viability of a product that influences the ability to attract funding during development. Furthermore, because revenues derive from the sale of products that have completed expensive and lengthy research programs, it is important to assess marketing issues at the outset of a research project in order to determine, as best as possible, that a profit-enabling market exists.

CRITERIA FOR SUCCESS

It is important to objectively evaluate factors supporting and challenging the profitable sale of biotechnology products and services early in their development. Three important criteria to consider in evaluating market conditions are:

- Freedom to operate
- Availability of technological factors
- Ability to generate a profit

Any criteria that are not satisfied upon initiation of a project will ultimately have to be satisfied in order for the project to succeed. For example, if a market is blocked due to lack of freedom to operate, licensing patents or designing around blocking patents can enable a company to serve a market and may also block competitors. Alternatively, if market access is impeded due to unavailability of technological factors or if existing methods are too expensive to permit profitable sales, performing the requisite R&D can facilitate access to the market, and protecting the R&D with patents or trade secrets can serve as a barrier to competitors.

If all criteria are met at the outset, as is often the case with generic drugs, then the barriers to entry are very low, unless a company has a competitive advantage. This advantage may be in the form of specialized expertise, intellectual property protection, or exclusive access to necessary tools or reagents.

FREEDOM TO OPERATE

To realize the potential of biotechnology inventions, companies require freedom to operate. Patents play an important role in defining freedom to operate. Competitor's patents may cover processes that limit the market opportunities that a company can target. Additionally, upstream and downstream patents can impede commercialization. For example, if a biotechnology product or service needs to integrate with a patented product or process, requires the use of a patented technique or product, or interferes with third party patents, agreements with the holders of the necessary patents will have to be reached. Alternatives to licensing requisite patents are engineering around patents, challenging them, or knowingly infringing them. Patent challenge is described in further detail in Chapter 7. In cases where patent challenge or engineering around a patent is impossible or prohibitively expensive, infringement may be the only available option. A significant downside of infringement is that damages may be trebled if a court can be convinced that a patent was deliberately infringed and the infringer is unable to demonstrate invalidity of the patent in question. Government regulations and public support or resistance can also influence freedom to operate; bans on stem cell research funding or transgenic crop planting, for example, have impeded biotechnology company operations.

TECHNOLOGICAL FACTORS

Biotechnology product development is a research-intensive endeavor. The bulk of effort is invested in either isolating and producing products, or in developing methods to isolate and produce products. In some cases the raw materials may already exist in a suitable form, whereas in others they must be produced or isolated from natural sources. Additionally, the process to produce a biotechnology product may or may not be known at the outset. Much of the effort in drug development is expended in identifying effective drugs and demonstrating their safety. In many gene therapy applications, for example, the gene that must be delivered for therapeutic effect is known and characterized at the outset. The challenge is to devise a method to deliver it in an effective manner. Conversely, the challenge for monoclonal antibody therapy is to produce and/or isolate appropriate therapeutic antibodies.

PROFITABILITY

Whereas the first two criteria, freedom to operate and availability of technological factors, are necessary to be able to produce a product, the ability to generate a profit is essential for commercial success. It is important to assess the projected cost of development, time to market, market size, and profit margins to effectively evaluate the merits of a project and select from multiple projects.

The process of satisfying the criteria to produce and profitably sell a biotechnology product or service requires three activities: research, development, and marketing. Research identifies potential lead compounds and technologies; development refines and characterizes these products and technologies in preparation for marketing; and, marketing identifies customer needs and develops methods to reach customers.

REGULATORY AND PUBLIC APPROVAL

A major concern for all biotechnology products is safety. The potential for unwanted effects on the environment and human health motivates government control and public awareness of applications of biotechnology. A unique aspect of the biotechnology industry is that any biotechnology product meant for human use, or which can interact with the environment, must receive regulatory approval.

Regulatory bodies are described in greater detail in Chapter 8. Briefly, the Food and Drug Administration (FDA) is responsible for ensuring that foods, drugs, and their manufacturing processes are safe for human consumption. The FDA also requires drugs to be proven effective and their labeling to be appropriate. The Department of Agriculture (USDA) regulates plant pests, plants, and veterinary biologics to ensure that they will not harm the environment or other plants and animals. The Environmental Protection Agency (EPA) ensures the safety of both chemically and biologically produced pesticides. A benefit of these regulatory burdens is that consumers are provided with objective determinations of the safety and efficacy of biotechnology products.

Beyond government regulations, public approval can also have a significant impact on marketability. Public aversion to

U.S. corn exports to European Union ($mm)

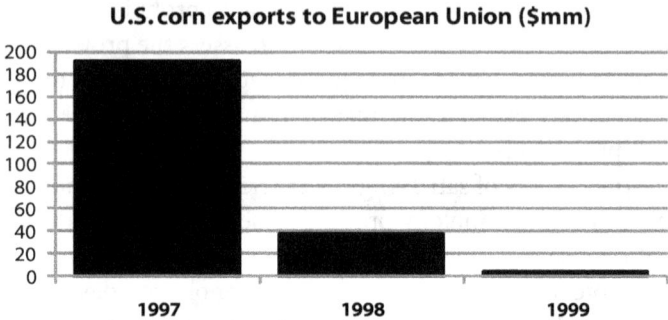

Figure 12.1 *Regulations affect sales: U.S. corn exports following partial EU ban*
Source: USDA Economic Research Service

genetically modified foods in Europe, partially attributed to a marketing campaign by a British supermarket aimed at displacing imported products, is implicated in the development of formal regulations in many EU nations banning the import of U.S. corn and other genetically modified foods. The impact of these regulations, shown in Figure 12.1, has been dramatic.

MARKET STRUCTURE AND MARKETING ENVIRONMENT

The purpose of market research is to develop an understanding of the market of a product, the potential customers, and the competitive environment.

Market research typically consists of two components: qualitative marketing research (also called primary market research) and quantitative market research (also called secondary market research). Primary market research involves the collection of original data from sources such as interviews, surveys, product tests, and focus groups. Secondary research is based on existing data from sources such as magazines and newspapers, scientific papers, industry reports, and other publications. Secondary research helps convert raw data from primary research into sales trends, demographic profiles, and business statistics that can help drive strategy.

Ideally, biotechnology companies should seek to serve a market that is large, growing, and underserved. It is important to consider all three criteria in evaluating markets. Large and growing markets permit generous, increasing, revenues, while underserved markets may feature low price elasticity and relatively easy customer acquisition. Asthma therapeutics, for example, represent a large and growing market, but the effectiveness of available medications challenges entrants to differentiate their products and offer a compelling advantage over existing alternatives. By contrast, treatments for geriatric ailments have a large and growing underserved market but are challenged by the ability of individuals and health care plans to support profitable sales.

MARKET SEGMENTATION AND TARGETING

Because most biotechnology products are developed for specific markets, targeting is an essential element of any biotechnology product marketing strategy. Targeted marketing is the alignment of marketing efforts for a product or service with its benefits to enhance sales. Market segmentation divides a market into distinct groups of buyers or decision makers. In targeted marketing, each segment of a market is evaluated for its commercial potential and a product's competitive advantages are identified and communicated to each segment. The assumption underlying targeted marketing is that focusing marketing efforts to where a product's unique benefits are valued most enhances commercial success.

The identification and selection of markets, permitting market segmentation and targeting, is facilitated by market research that defines market structures and environments. Once favorable markets are identified, the next objective is to critically evaluate which ones are preferable.

Markets should be measurable, accessible, actionable, and substantial. Measuring the size and value of a market permits an evaluation of the potential for revenue and comparison with other markets. Markets must also be accessible and actionable, because a market that cannot be reached cannot deliver revenues. Finally, a market must be large and profitable enough to warrant targeting. Small biotechnology companies may license shelved drug leads from larger companies because of their ability to profitably serve smaller markets (for an example of drug development for markets without profit potential, see Box *Nonprofit drug development* in Chapter 11).

MARKET SIZE

When assessing the size of a market it is not sufficient to simply cite the number of people with a given condition, because it is unlikely that all, or even many, of these individuals will take a given drug. Figure 12.2 shows the progression from determining disease prevalence (the number of people who have a disease) to measuring the size of a target market. For example, while the prevalence of hypertension in the United States is 30 percent—approximately 100 million have hypertension—this figure does not represent a target market. Most people are unaware that they have hypertension, and many who are aware do not seek treatment. Even among those who seek treatment, many fail to fill their prescriptions. So, a company producing a hypertension treatment cannot hope to sell their product to 100 million people in the U.S.

To determine the size of the target market, one must consider how many patients are aware that they have a condition, see a physician, obtain a prescription (for the appropriate disease) and fill that prescription. In 2003 there were more than 35 million physician office visits for hypertension, representing a presentation rate of roughly one-third.[1] This figure must still be discounted by the number of repeat visits (how many unique individuals are represented in these 35 million visits), the number of individuals who did not receive prescriptions for hypertension medications, and those who did not fill their prescriptions.

Market sizes may also be a source of contention in licensing, partnership, or funding and acquisition discussions. The

1 Centers for Disease Control

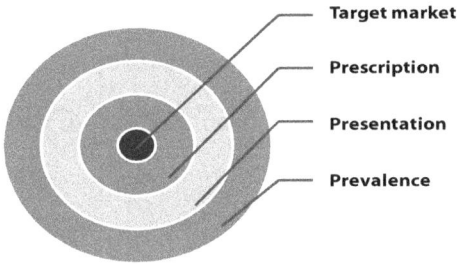

Figure 12.2 *Measuring target market size*

potential for market projections to influence the price of an individual product or entire company may lead to disagreements over calculations and methodologies.

REIMBURSEMENT

Reimbursement is a critical element in commercializing drugs. Patients seldom pay the full price of the drugs which they are prescribed. Third-party payers such as health insurers and government programs usually cover all most of the cost. Failure to consider reimbursement can prevent commercialization of an approved product. FDA approval alone is not sufficient to ensure sales; payers must also be willing to reimburse patients for a drug.

CONSUMERS, GATEKEEPERS, AND PAYERS

The ethical drug market is a complex ensemble of consumers, gatekeepers, and payers. The parties responsible for selecting drugs, using them, and paying for them each have different motivations and needs. The challenge for firms serving this market is to identify the various types of customers and meet their specific needs.

Figure 12.3 shows a simplified model of the flows of money and drugs in the ethical drug market. In this model, drugs are supplied to the pharmacy from the manufacturer, often through a series of wholesalers and distributors. Patients are

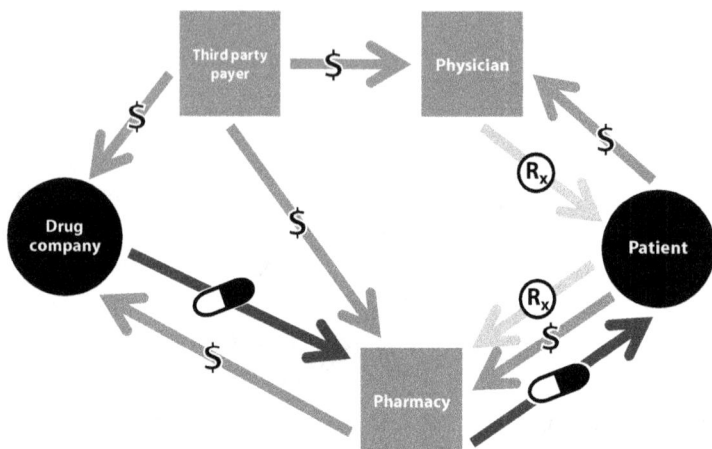

Figure 12.3 *Simplified drug market model*

given prescriptions by a physician and take the prescriptions to a pharmacy where a pharmacist dispenses the requested drug. Underlying this series of drug and prescription transactions is a network of entities and incentive mechanisms that ultimately fills prescriptions and compensates all the players. Physicians are compensated by third-party payers (insurers, Medicare, Medicaid, etc.) and co-payments from patients. Pharmacies are also compensated by third-party payers and co-payments from patients.

Third-party payers exert influence over prescription patterns by dictating which drugs are reimbursed, which drugs require prior authorization for prescription, which substitutes should be applied, and how much co-payment is required. A drug's pharmacoeconomic profile, or cost-benefit ratio, can have a profound effect on prescription patterns. Third party payers are motivated to reduce their healthcare expenditures. Drugs that can shorten hospital stays, substitute for expensive treatments, prevent diseases which are expensive to treat, or otherwise reduce the cost of health plan administration are therefore able to command premium prices.

Physicians are the ultimate gatekeepers, deciding which

drugs to prescribe. The physician's primary consideration is the expected clinical improvement of the patient. In addition to safety and efficacy, physicians must also consider factors such as administration and monitoring. Drugs which place a burden on healthcare professionals may be reimbursed at the same rate as simpler alternatives, creating a financial disincentive to their prescription. Side effects and tolerability place a burden on a physician's time. The need to monitor side effects and clinical progress through repeated visits or laboratory testing further increase this burden. Accordingly, a drug with a high maintenance regimen or even a low incidence of particularly dangerous side effects may be less preferable to a lower-maintenance drug with a more predictable side effect profile (see the examples of Herceptin and Aczone in Box *Personalized medicine and drug sales* in Chapter 6).

Patients also have an impact on drug sales. To visit a physician and obtain a prescription, patients either need to suffer an incident which brings them to a hospital, or otherwise realize they have a condition for which a treatment may exist and feel compelled to seek treatment. Patient education programs are commonly used to encourage potential patients to consult physicians regarding treatment options.

LIFECYCLE MANAGEMENT AND MARKETING MIX

Traditional product lifecycles are often drawn as bell-shaped curves (see Figure 12.4, top panel). For innovative products such as patented drugs or those facing obsolescence when improved alternatives emerge (such as bioinformatics software and databases, or research tools), product adoption rates can be relatively faster and decline rates more precipitous (see Figure 12.4, bottom panel), although the absence of clear guidance on generic biologic approval can prevent price competition following patent expiration.

As a product goes through the life cycle of development; introduction; growth; maturity; and, decline and/or patent ex-

a) Traditional product lifecycle

sales

| Development | Introduction | Growth | Maturity | Decline |

b) Biotechnology product lifecycle

sales

| Development | Introduction | Growth | Maturity | Decline / patent expiration |

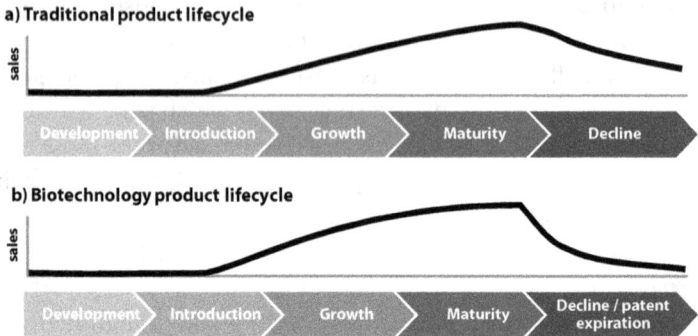

Figure 12.4 *Biotechnology product lifecycle*

piration, there are different strategies that influence marketing activities for the product. For example, for discovery through growth, intense marketing is required to get the most out of the investment in the product. As a product matures and declines, less is spent and higher returns are sought on the existing spend. Usually when a product goes off patent, marketing spend and sales support are withdrawn as alternatives (e.g. generic drugs) enter the market and capture market share based on discounted price strategies. Strategies to delay or offset market decline include patent life extensions (described in Chapter 7) and R&D strategies to develop brand extensions (described in Chapter 11).

The distinct lifecycle stages also influence the set of marketing tools and tactics, also known as the marketing mix, used to optimize the distribution of the marketing budget to maximize sales. Whereas early-stage activities may focus on product positioning and reinforcement of value messages, later-stage activities may favor brand awareness to preempt generic and encourage a shift to follow-on products.

Chapter 13

Managing Biotechnology

> Marketing and innovation are the two chief functions
> of business. You get paid for creating a customer,
> which is marketing. And you get paid for creating a
> new dimension of performance, which is innovation.
> Everything else is a cost center.
> *Peter Drucker*

As described in the opening quotation by the late Peter
Drucker, innovation and marketing are the two chief
functions of business. The basic role of management,
therefore, is to guide innovation and marketing, and to assemble and maintain the supportive elements which facilitate them.
This chapter integrates themes from the previous chapters to
provide perspectives on managing biotechnology companies.

While biotechnology companies are appropriately described
as R&D-intensive, it is important to recognize that a number
of other operations are required to enable the commercialization and profitable sale of innovations. Figure 13.1 presents a
cost analysis of the pharmaceutical industry as a general model
of expenditures in a mature biotechnology companies. Several
of these expenditure categories can be affected by new innovations. Developing products using biotechnology techniques can
directly affect the cost of R&D and manufacturing—biotechnology drugs tend to cost more to develop and manufacture
than pharmaceutical drugs. Secondary effects stemming from
the use of biotechnology, such as defining markets through the

use of functional genomics or filling unmet market needs with innovative products, can positively impact marketing and sales operations, reducing costs and increasing profits.

STARTING UP

Biotechnology companies form to develop and exploit new technologies. Conditions in which start-ups thrive are characterized by disruptive changes that permit the development of products and services to satisfy unmet market needs. Biotechnology allows the development of novel products, cheaper and more efficient manufacture of existing products, and development of more effective and refined versions of existing products.

A start-up's goal is to attain sufficient funding to reach milestones that will raise its valuation and lead to either further financing or liquidity. In order to succeed, a company must have a unique competitive advantage that positions it to succeed in its endeavors.

A company's fundamentals change as it matures. While business plan formulation is essential early-on to outline long-term goals, business plans inevitably change as a company ma-

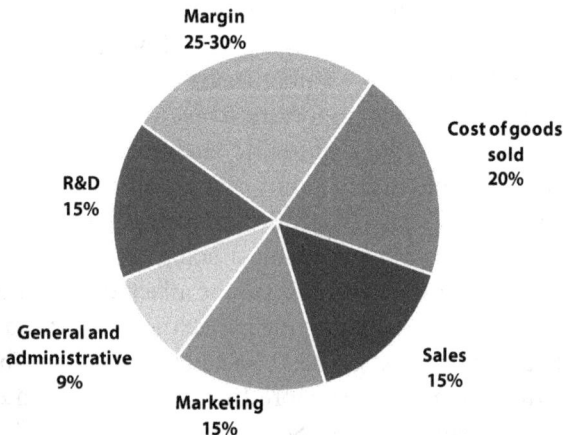

Figure 13.1 *Pharmaceutical industry cost analysis*
Source: Datamonitor, 1999

Figure 13.2 *Producing and selling biotechnology products and services*

tures. Research and development are also very important at early stages, enabling the development of marketable products and a revenue stream, or at least the assurance of one.

Once a product has been developed, management needs to establish manufacturing and commercialization abilities. Early marketing and sales agreements provide income and will have a long-term impact on a company's development. To ensure continued success, a balance must be established between research and commercialization; a broad pipeline must be complemented by manufacturing and sales abilities that will result in commercial success.

FAIL FAST

A recent study from the Tufts Center for the Study of Drug Development found significant differences between pharmaceutical companies in the time required to develop drugs.[1] The five fastest companies between 1994 and 2005—Bayer, AstraZeneca, Allergan, Boehringer Ingelheim and Merck—had as much as a 17 month speed advantage over average performers. A follow-up study of clinical development executives at the top-performing firms identified four key activities associated with increased performance:

- Enterprise-wide adoption of e-clinical technology solutions
- High usage of contract clinical service providers
- Active interaction with regulatory agencies
- Effective management and prioritization of resources,

---

1 Kaitin, K.I. *et al.* Fastest drug developers consistently best peers on key performance metrics. *Tufts Center for the Study of Drug Development Impact Report*, 2006. 8:5.

including earlier termination of poor projects[2]

The early termination of poor projects was echoed in the Tufts CSDD report, which found that fastest drug developers terminated the majority of cancelled projects in Phase I, versus the slowest drug developers who terminated the majority of cancelled projects in Phase II. This strategy can also deliver great cost savings, as the cost of clinical trials grows progressively with each phase.

INTELLECTUAL PROPERTY PROTECTION

Intellectual property can form a barrier to entry of competitors and attract investors and partners. Because poor security of intellectual property can threaten a company's competitiveness, discourage investors and partners, and put a company's very survival at risk, the quality of a company's intellectual property protection is central to its prospects for success. The sections *Patent Challenge* and *Extending Patent Protection* in Chapter 7 describe strategies to protect intellectual property and maximize the lifespan of patents.

Intellectual property protection is of central importance in biotechnology. The primary role of intellectual property protection is to enable a company to exclude competitors from a market. While considerable effort is generally required to produce a new biotechnology product, copying is often much simpler. The sophistication and widespread availability of biotechnology research tools limits the ability of trade secrets to protect inventions, due to the likelihood of reverse-engineering by competitors. Because of the importance of maintaining a competitive advantage, the scope of patent protection and market exclusivity have an important influence on which products biotechnology companies will develop. Patents and other forms of intellectual property protection are discussed in greater detail in Chapter 7. In addition to securing future revenue streams, the

---

2 Frantz, S. Study reveals secrets to faster drug development. *Nature Reviews Drug Discovery*, 2006. 5:883.

quality of a company's intellectual property is a key consideration of investors, partners, and acquirers, impacting the value of a company.

MANAGEMENT CHANGES WITH GROWTH

A company's management team is the ultimate source of leadership in strategy, implementation, and financing. In a young company the founding entrepreneur and financial sponsor may play some or all of these roles. As a company grows, it becomes necessary to add specialists and dedicated staff members to relieve the burden on founders, enabling them to focus on the activities for which they are best suited and bringing much-needed operational experience to a maturing company.

A difficult time in the growth of any company is the point where founders must step down and allow others to manage the company they have built. In some cases these transitions may be initiated by financiers, and in others they may be initiated by founders seeking to start new ventures, or to free themselves of the burden of running a biotechnology company.

While founders may manage diverse activities such as R&D, partnerships, and human resources, in mature companies these roles require specialized skills, and the people who perform these roles may come from fields unrelated to biotechnology where they have developed their talents. One of the downsides of failing to recruit specialized talent is that the founding management team may be poorly positioned to function effectively in the market in which they operate. Additional sources of domain expertise are the board of directors and scientific advisory board. These esteemed industry leaders and experts can be tapped to fill management gaps in growing firms.

V

Conclusion

Once you understand the various legal, regulatory, political, commercial, and scientific factors that define, enable, and constrain the biotechnology industry, it is possible to apply this knowledge in many ways. This conclusion presents a framework for investing in biotechnology, and guidance in career development.

Chapter 14

Investing

If you don't know jewelry, know the jeweler.
Warren Buffet

CAVEAT EMPTOR

Less than an hour after shares for Genentech's initial public offering (IPO) opened at $35 in 1980, the price had appreciated to $88, making for one of the largest stock run-ups ever, and casting biotechnology in investor's minds for decades to come.

Investors are still drawn to biotechnology today because of the high profit margins, years of patent-protected sales, and the ability to address pressing needs in growing markets. However, a measure of perspective must be applied in evaluating the potentials of investing in biotechnology.

Most biotechnology companies make for poor investments. Relatively few of the multitude of biotechnology companies eagerly developing innovative products to address lucrative markets will ultimately succeed. Of those few that do succeed, only some will be profitable, and only some of those will deliver outstanding returns. What makes investing in biotechnology so compelling is the historical ability of just a few companies to deliver such outstanding returns that they effectively prop up the entire sector.

A ten year follow-up analysis of the 41 biotechnology companies that went public in 1995 and 1996 found that:

- 25 were still independent
- 4 had gone out of business
- 12 had been acquired
- 7 had been acquired at a loss
- 6 were profitable
- 15 had a share price higher than the close on their first day of trading[1]

This basic analysis highlights an important trend. Even among the select biotechnology companies that are able to mature into public entities, most fail to deliver positive returns over an extended period of time. According to Eaton Vance Worldwide Health Sciences Fund manager Samuel Isaly, only 40 of the 1,000 biotechnology companies that have gone public in the history of the industry have ever attained profitability.[2]

INVESTING IN BIOTECHNOLOGY

Despite the numerous challenges of investing in biotechnology companies, investors have the ability to participate in the development of products that benefit humanity and may also realize outstanding financial returns. To invest profitably in biotechnology companies, one must appreciate the influences of scientific, legal, regulatory, political, and commercial factors. While there are a variety of strategies for investing in public markets, in developing a biotechnology investment strategy it is vital to consider the unique characteristics of biotechnology companies and the challenges they face.

Biotechnology companies are very research and development-intensive. Start-up companies are especially risky investments because they have high burn rates and sell on the promise

1 Travers, C. Grading old-school biotech. *Motley Fool*, January 25, 2005. http://www.fool.com/news/commentary/2005/commentary05012506.htm

2 Jacobs, T. Great company, bad stock. *Nature Biotechnology*, 2005. 23(2):173.

of future profits. For mature and start-up companies alike, the success of any individual project is far from certain. Research projects can fail at any stage for predictable or unpredictable reasons. The great uncertainty of whether or not product development will be successful challenges the formulation of financial projections, limiting the predictive ability of traditional valuation methods.

Faced with the inherent uncertainty of biotechnology development, investors must learn to accept the risks associated with investing in biotechnology companies. This book has described the scientific, legal, regulatory, political, and commercial factors specific to biotechnology companies. Understanding the importance of all of these subjects in biotechnology company development allows the educated investor to objectively assess an investment opportunity.

While identifying successful companies enables investors to profit from breakthroughs in biotechnology, poor stock selection is unlikely to yield any gains. An investor who does not appreciate the unique challenges and opportunities of biotechnology research is better off purchasing general sector mutual funds, investing in the companies that stand to indirectly benefit from biotechnology, or avoiding biotechnology investments altogether.

STRATEGY

A fundamental difference between investing in biotechnology companies and investing in more traditional companies is that traditional analysis methods are less able to predict success in biotechnology. This is particularly relevant for companies without products on the market. Biotechnology product development is fraught with unexpected failures. Projects may fail for any number of reasons that could not be predicted at the outset. Share prices, especially those of smaller companies, may fluctuate significantly in response to project development progress. Significant R&D successes can see stock prices appreciate rapidly; disappointing results can lead to precipitous price drops.

It is necessary to consider multiple factors to minimize investment risk and predict the future prospects of biotechnology companies. Highly-focused analysis of revenues and expenditures, for example, is of limited value for an early-stage biotechnology company with characteristic high R&D investments and the future potential of substantial income. It is important to assess the probability of achieving a future revenue stream in addition to measuring the possible magnitude of future revenues.

FUNDAMENTALS ANALYSIS

A general survey of desirable characteristics in biotechnology companies reveals the following criteria for promising investments:

Corporate Qualities
- Maturity
- Business model
- Experienced management
- Competitive advantage
- Institutional support
- Favorable financial analysis
- Sufficient funding to fund operations for several years

Product Qualities
- Successful products on the market
- Numerous products in development
- Products that target large or underserved markets

The above criteria are subjective, requiring a fair amount of personal judgment in evaluating the quality of a company's strengths in each area. Furthermore, the above criteria do not necessarily lead to success but can identify likely failures. A company with excellent fundamentals may fail, but a company with poor fundamentals cannot be successful. It is therefore important to not obsess over individual measures, using the criteria instead as a framework upon which to structure an overall assessment.

CORPORATE QUALITIES

Maturity

The maturity of a biotechnology company is an important consideration in assessing corporate strength and stability. Summarizing the grouping of biotechnology companies by maturity and stability as presented in Chapter 9, biotechnology companies can be roughly grouped into three categories. Established large-cap firms with positive revenue streams are labeled mature. Those with strong fundamentals and excellent prospects for near-term profitability are labeled promising, and the remaining biotechnology companies, those without near-term certainty of sustainable profits, are labeled emergent.

Briefly, mature companies are the most amenable to traditional financial analysis. In addition to a mature company's historical record and current financial position, it is also important to consider non-financial factors such as the business model, management track record, quality and quantity of partnerships, and apparent strength of research efforts in order to project future performance. Mature companies usually have a number of successful products on the market as well as proven ability to maximize revenues and protect market share.

Promising companies are more difficult to objectively evaluate than mature companies. The future prospects of these firms often depend on events such as success of key clinical trials, the outcomes of which cannot be reasonably predicted. Evaluating revenue streams of promising companies is less predictive of success than it is for mature firms, because the expectation is that the revenues of promising companies will grow significantly in the future. A more important factor in assessing promising companies is financial health. In order to succeed, a promising company must have sufficient finances, or the potential to raise sufficient finances, to fund operations and growth in order to develop into a mature firm. Other important considerations are the quality of non-financial parameters such as the business model, management expertise, quality and quantity of partnerships, and strength of research efforts. While it is difficult to predict revenues for sales of yet-to-be developed novel products, it may be productive to project revenues and use these figures in a traditional financial analysis to determine if a company's current stock price already reflects positive future expectations. While investing in promising companies involves significant risk, emergent companies present even more risk. Financial analysis of an emergent firm using projected revenues may be an effective way to determine if a company's value exceeds even optimistic revenue projections, but the uncertainty of revenues for these firms limits the utility of this exercise. Furthermore, emergent company business plans, business models, and target markets may change significantly prior to development of a stable revenue stream. The unpredictability of future research directions and revenue possibilities means that the most effective way to invest in emergent companies is to accept a large component of risk and look at fundamentals that provide a strong research infrastructure such as access to capital, quality of the management team, quality of partnerships, and aggressiveness of research efforts.

BUSINESS MODEL

One of former Fidelity Magellan Fund manager Peter Lynch's investing principles is, "never invest in any idea you can't illustrate with a crayon." It is vitally important to understand, even at a superficial level, how a company is going to make money. It is easy to be enticed by innovative technologies or large markets, but one must rationalize how an investment can increase in value. Promising companies may employ innovative technologies or seek to serve lucrative markets, but innovative technologies and products do not create revenues; sales do.

Management Team

The quality of a company's management team is arguably the best predictor of success. This is especially true for young companies. Individual research projects may succeed or fail, but it is ultimately up to management to secure resources to enable research, to guide research towards profitable ends, and to facilitate the profitable sale of developed products and services. Managers with a history of success are likely to succeed again.

A company's managers should have experience in general skills such as managing collaborations and securing financing, as well as in areas pertinent to the specific development plan such as unique regulatory requirements or development hurdles. Past experience dealing with the issues that a company is likely to face prepares management to predict and swiftly resolve problems.

Some history of failure among managers is not necessarily an indication of unsuitability. Ironically, failure at a previous firm can be a more informative experience than success. A share of failure keeps people humble. Failures also teach people to do things differently and can provide insight into which actions should and should not be taken in unfamiliar situations. One of the factors for the successes of American entrepreneurs in biotechnology and other domains is attributed to the relative willingness among investors to back people who have failed in previous ventures.

Competitive Advantage

As described in Chapter 7, biotechnology companies require a competitive advantage to prevent competitors from capitalizing on the efforts of pioneers and denying them the ability to recover investments in research and development.

The number of competitors and the developmental maturity of competing products are important considerations in assessing the value of a company's products. The safety, efficacy, and price of competing products are also important factors in predicting the impact of competition on market share.

For biotechnology companies, patents are the most commonly used means to secure a competitive advantage. Patents grant innovators 20 years of exclusive rights to exclude others from making, using, offering for sale, or selling an invention. Patenting an invention requires the innovator to disclose the best means to practice an invention, which can facilitate the emergence of competitors upon patent expiration or invalidation. Key technologies should be protected by multiple patents to protect against invalidation of a single patent and to cover as broad an application area as possible. Beyond simply restricting discrete applications, an additional benefit of employing multiple patents is that they can also convey a willingness to aggressively defend intellectual property. In one example, Affymetrix used more than 400 patents to protect its microarray technology, raising a significant legal and scientific barrier to competitors. The great cost of legal expenses to simply determine freedom to operate in the microarray space likely dissuaded many potential competitors.

A law degree or Ph.D. is not necessary for a general assessment of a company's patent strength; examining a company's history can be telling. A company with a history of winning patent decisions is likely to continue to succeed in protecting its intellectual property. Furthermore, an established reputation for securing favorable judgements can discourage infringers and encourage favorable out-of-court settlements.

While patents are not the only way to secure a competitive advantage in biotechnology, they are the most common. Additional forms of competitive advantage include trade secrets, restricted access to superior sales forces, distribution networks, and lucrative partnerships. In the final analysis, it is important to assess the ability of any competitive advantage to exclude competitors and protect profits.

INSTITUTIONAL SUPPORT

Partnerships provide both cash and endorsement, and are of great importance in the biotechnology industry. Partnerships with industry leaders indicate that knowledgeable and capable industry insiders endorse a junior partner's scientific and commercial possibilities. Because large companies and venture capitalists often respectively partner with, and fund, multiple companies with the knowledge that they can cut poor performers loose, it is important to consider institutional support in the context of other company fundamentals.

In assessing the quality of a partnership one should look for cash and rights distributions. The magnitude of upfront payments, milestone payments, downstream royalties, and co-promotion rights are all measures of the value of a partnership. Large cash commitments, especially upfront payments, indicate strong support. Furthermore, the obligations of each partner (e.g., does the junior partner have to simply identify leads, or must they produce a drug that passes clinical trials?) indicate the relative competitive strengths of each partner. The relative stake of each partner dictates how any profits will be distributed.

Just as partnerships with esteemed firms represent an assessment of a company's value, the quality of the money behind a start-up is another measure of industry insiders' assessment of a start-up's prospects. Backing from a top-flight venture capital firm with experience in biotechnology is an excellent indication that some very knowledgeable people think that a company's prospects are good. One way to assess the quality of a financial backer is to look at their track record. A backer with a history of successful investments in biotechnology is likely to experience future success. Backing from major corporations with relevant industry experience is another positive indicator.

FINANCIAL ANALYSIS

While the other biotechnology company evaluation criteria in this section describe methods to assess whether or not a company is likely to succeed in product development, financial analysis is also important because it provides information on whether a stock's value represents the value of a company. A company may have strong prospects, but it is also important to determine if the stock's value represents, or even exceeds, expectations. Because many companies lack significant revenues, financial assessment of biotechnology companies is challenging. Even those companies with sizable revenues often have uncertain futures.

A number of valuation models have been proposed over the span of decades to determine the value of biotechnology companies. Models for success in biotechnology have considered factors such as the number of Ph.D.s employed, market forecasts, cash flows, and the ratio of R&D expenditures to earnings. Unfortunately, a method to effectively reduce biotechnology development and market potential to mathematical formulas does not yet exist. Selecting peer groups, projecting revenues, and assigning development risks are subjective measures. Calculations based on these figures are likewise subjective. Financial analysis can provide a useful metric for biotechnology, but it requires careful integration with non-financial parameters to increase relevance.

Peer comparison is a method used to determine if a given company's share price is relatively more or less expensive than other similar companies. Peer comparison can also be used to examine whether certain biotechnology sectors are relatively over- or under-valued relative to other sectors. The simplest tool for peer comparison is calculation of the ratio of stock price to annual earnings. This so-called price to earnings (P/E) ratio is calculated by dividing a company's current share price by the earnings per share for the previous twelve months. P/E ratio calculations are of little relevance for companies without earnings.

There are a number of other peer comparison measures which are more sophisticated than P/E ratios, but they all share a common weakness. The assembly of a peer group requires a certain amount of subjectivity. Furthermore, biotechnology peers rarely develop similar products for similar markets, meaning that peer differences are expected and will be based on non-financial attributes. Most importantly, peer comparison only yields relative valuations. A company may be undervalued or overvalued relative to its peers, but if it is in a misvalued sector its share price can nonetheless fall or rise in response to market sentiments rather than its intrinsic value.

Burn Rate

For many biotechnology companies, product launch and profits are often many years away and require significant financial investment. Access to cash is an important factor in enabling a company to complete development, making a company's rate of spending, or burn rate, an important measure. Stock analysts often look for companies with a minimum of two years' cash reserves. This time span should either permit the completion of milestones that facilitate future funding, or lead to profitable revenue streams.

PRODUCT QUALITIES

Products in Development / on the Market

Products on market are not only a source of revenue, they are also a testament to management's ability to guide a product through development and commercialization. Companies with product revenues can use this incoming cash flow to maintain operations and to fund future development. The ability to draw upon internal resources also means that these companies will likely be able to maintain greater independence in future product development and retain a greater share of profits.

For companies without successful products, those with products in late stages of development present less risk than those with products in early stages. The process of clearing a potential drug for clinical trials and proceeding through the three phases of clinical trials to demonstrate safety and efficacy is very challenging and unpredictable. The more advanced a company's products are in this process, the lower the risk of significant setbacks or failure.

In judging development progress, it is important to consider the source of a report. Many journalists struggle to understand the fundamentals of biotechnology research and may inadvertently make errors in reporting. One should examine whether a report comes from a respected newspaper, magazine, or broadcast, and if it is produced by individuals with proven expertise. Furthermore, while the FDA regulates press releases about drugs that have been approved, there is relatively little oversight of press releases about experimental drugs. Reports based on single Phase I trials, for example, may project enormous market potential well before the safety and effectiveness of a drug is established.

PIPELINE DIVERSIFICATION

Biotechnology product development is fraught with unexpected hurdles, setbacks, and failures. Accordingly, it is imperative that companies have multiple products in development to provide alternatives should individual research lines fail. Multiple products are essential to support long-term growth.

A company's development pipeline should ideally be sufficiently broad to address markets with good revenue potentials and to provide a measure of stock price stability. Excessive diversification can be as detrimental as a lack of diversification. A lack of focus in research projects demonstrates poor management. Furthermore, a company pursuing too many unrelated objectives may be unable to focus sufficient resources to overcome significant development complications or enable commercial success.

LARGE OR UNDERSERVED MARKETS

Significant long-term value can be derived from products which are sold to a broad customer base. This is the reason why drugs, food products, and health services are common applications for biotechnology development. Another factor in assessing the quality of a product is evaluating how pressing a need it serves and how frequently it is likely to be purchased. For example, a drug serving a chronic life-threatening condition is likely to generate greater revenues than an infrequently used or non-essential drug.

Products in development should have defined markets and clear advantages over any existing alternatives. It is important to not overestimate the potential of poorly defined products serving defined markets or defined products serving poorly defined markets. Celera faced a crisis after completing its goal of producing a rough draft of the human genome when it found that the market for genomic information was insufficient to support desired growth rates. This misjudgment required Celera to divest itself of its genomics business to focus on drug development, before shifting focus again to creating molecular diagnostic tests.

It is also important to assess the barriers to successful development:

- How well is a problem defined?
- How refined are the tools needed to solve the problem?
- How much research and development is necessary?
- If a company is developing applications of a promising technology that has eluded others for years, then what are the odds of success? What new tools or techniques are they using?
- Has a recent scientific or technological development created new opportunities? Is there a way to vet the potential of this recent advance?

A poorly defined problem introduces the possibility of large, unpredictable, capital requirements. The availability of appropriate tools to develop an application reduces the uncertainty of capital and time requirements and allows for defined and predictable milestones. It can be difficult to determine which new technologies are best able to reduce uncertainty, but one positive metric is the entry of new players using a new technology. Applications with great potential will likely attract multiple developers. Revolutionary technologies such as gene splicing, monoclonal antibodies, gene therapy, and RNA interference all led to the emergence of new companies focused on leveraging them. Not all of these technologies delivered on their initial promise, but weak adoption can be a good sign that a technology is unlikely to deliver great returns. Beyond the qualities of individual products, it is also important to consider a company's prospects for future development. For example, a company which has been able to consistently produce moderately valuable products may fare better in the long-term than a company which infrequently produces more valuable products.

INVESTING ON TRENDS

In addition to assessing the fundamentals of companies, it is important to consider market trends. The volatility of biotechnology stocks challenges the interpretation of trends and the application of this information, but there are some important factors to consider.

When market support is strong, undeserving companies often see their share prices rise as they are associated with market leaders. When market support weakens, even good companies will see their prices unfairly depressed. In the long run, companies with strong fundamentals are likely to succeed, whereas companies with poor fundamentals may be acquired, liquidated, or stagnate. This reality underscores the importance for long-term investors to evaluate fundamentals.

Short-term investors who purchase company shares primarily in the hopes of near-term FDA approval should consider the possibility of a share price collapse should a drug be found unapprovable. By investing early, investors may be able to secure relatively greater returns, but they also face significant risk. MSN Money examined the performance of companies following FDA review of sixty biotechnology drugs from 1998 through 2000. They found that 55 percent of the tracked stocks were trading higher thirty days after approval, with an average gain of 7 percent. Sixty days after approval the average gain was 13.4 percent, with 58 percent of the stocks trading higher. Ninety days post-approval 60 percent of stocks were trading higher; an average gain of 12.6 percent.[3] Conversely, shares can

3 Niederhoffer, V., Kenner, L. Biotech: an investing frontier for risk-takers. *MSN Money*, July 5, 2001.

fall 70 percent or more in a single day following failure of a key drug to receive FDA approval. While this historical example is not indicative of future market performance, the potential to average a greater than 13 percent gain by buying drug stocks on the day of approval and holding them for two to three months represents a simple and potentially lucrative way to invest in biotechnology.

CONCLUSION

Any model to determine the value of a biotechnology company or likelihood of commercial success relies on a certain degree of personal opinion, informed or otherwise. It is important to be aware of the relative contributions of measurable and unmeasurable factors in assessing the merits of individual companies. Following meticulous financial analysis, for example, a company may be deemed undervalued or overvalued based solely upon which other companies it is compared against. Similarly, overestimating or underestimating the future potential of a product can effectively make all other measures insignificant. It is therefore necessary to consider whether the sum of measures forms a consensus; investment decisions should not be overly reliant upon individual measures.

Although assessment of the criteria described above cannot identify certain long-term winners, it can effectively identify losers. Consider a company that fails in multiple measurement criteria—poor cash position, an uninspiring management team, and weak technology—such a company is almost certain to fail. Even in the unlikely event that a lucrative breakthrough emerges, a poorly equipped company will not able to capitalize on it.

Investors should take the time to learn about the unique aspects of biotechnology companies and reflect on them in making investment decisions. Investors who are unable or unwilling to make this commitment may fare better in selecting alternative methods to capitalize on growth in this sector. While small biotechnology companies may develop lucrative products, they are also likely to form research, production, or marketing alliances with industry leaders. Investing in the biotechnology and pharmaceutical leaders that share in the success of smaller firms is a relatively safer way to invest in biotechnology.

Another alternative is investment in sector mutual funds. These investment vehicles tend to roughly mirror the rise and fall of a whole sector and permit investment in biotechnology without the volatility associated with individual companies. Mutual funds range in investment focus from specific sub-categories of biotechnology such as genomics to broad funds covering healthcare in general. As with individual stocks, investors should only invest in mutual funds which cover areas that they understand. Investing in a niche fund without understanding the future potential of that niche is likely to offer the same poor returns as investing in a company without understanding its commercial prospects.

Chapter 15
Career Development

Choose a job you love and you will never work a day in your life.
Confucius

Although the scientific aspects of biotechnology may seem daunting to outsiders, biotechnology companies employ many of the same job functions as other firms. In addition to researchers and technicians, biotechnology companies also employ receptionists, lawyers, engineers, janitors, salespeople, accountants, and a host of other business professionals.

While a Ph.D. in science is not a necessity for a career in biotechnology, it is important to understand topics relevant to one's role and to be able to communicate effectively with others. A computer programmer on a bioinformatics project, for example, must know enough biology to communicate with biological scientists. Salespeople should understand the applications of their wares, permitting them to advise customers of useful products. While there are numerous opportunities outside of research and development for individuals without advanced degrees in science, the nature of R&D requires individuals with refined knowledge and skills.

EVALUATING POTENTIAL EMPLOYERS

The performance of a company and the sector it is in influences the job stability and role of its employees. For example, the genomics sector cut 1,500 jobs from January 2001 through June 2002 following loss of public market support.

Job candidates should spend at least as much time selecting an employer as one might spend selecting a stock for investment. Candidates should look at the quality and experience of a company's management team, technology of products being developed, cash position and soundness of financing, quality of investors, competitive advantage, market need, and competition. Asking about the tenure of the management team as well as future hiring plans can indicate the stability of a firm and suggest whether a hiring decision is based on growth or replacing lost functions. For start-ups, where answers to many of these questions are unavailable or unknown, discussions of a company's burn rate and plans for future financing can indicate job stability and the likelihood of being redirected to different projects. Asking about a start-up's exit strategy and time frame can likewise reveal information about potential job stability and impending transformative events.

Other important elements include benefits such as retirement savings plans and health insurance. Companies may also offer employees stock options to offset lower salaries or to encourage continued tenure. It is important to consider how and when these options can be liquidated. Stock options in private companies may only have value if and when the company becomes publicly traded, and in public companies the stock options may likewise have limited value if their purchase price exceeds the market price (due to the options being issued at a time when the market price was higher).

While companies are unlikely to divulge complete details to job candidates, asking the right questions can reveal corporate strengths and weaknesses, and can convey an understanding of the fundamentals of commercial biotechnology to interviewers.

JOB DESCRIPTIONS

RESEARCH ROLES

Individuals with advanced scientific knowledge and abilities conduct and guide research. The role and responsibility of researchers varies with their individual skills and expertise.

Laboratory technicians perform various maintenance tasks in laboratories and may perform experiments as well. Most laboratory technicians have master's degrees, but opportunities are available for those with bachelor's degrees or high school diplomas. The functions performed by technicians range from maintaining stocks of reagents and research supplies to performing or supervising routine operations and performing supervised research experiments. Important skills for research technicians are communications and precise record keeping.

Opportunities for individuals with Ph.D.s, who may or may not have post-doctoral training, vary from research team members to core facility managers. Initial posts for young scientists are as members of research teams. After developing specialized expertise, seasoned scientists may lead research teams, plan and manage multiple research projects, and run labs.

The scientists who work in biotechnology companies generally have backgrounds in biology, chemistry, and medicine. A scientist's background influences which roles they will perform. Chemists, for example, will most likely find themselves engaged in early drug discovery. Developing potential drugs in preparation for clinical trials will likely involve pharmacologists, and running clinical trials requires physicians.

Non-research roles such as management positions, intellectual property management, and consulting require intimate knowledge of scientific fundamentals and research dynamics. Although individuals in these fields do not necessarily require experience in the specific field of research they are working in, advanced training such as a master's or Ph.D. degree in a related field is a great asset.

NON-RESEARCH ROLES

Aside from research and development, biotechnology companies perform many of the same operations as other companies. Directors and managers, for example, generally have business or management degrees (often in addition to scientific doctoral degrees). The central role of funding and financial management in biotechnology establishes a demand for individuals with proven financial expertise. Likewise, communications and human resources professionals are also needed for their specialized abilities.

Because of the importance of intellectual property protection in biotechnology, lawyers are needed to compose and prosecute patents and assist in the collection and evaluation of competitive intelligence. Marketing and sales experts are needed to study and develop markets and ultimately enable the delivery of products to consumers. Furthermore, the potential for substantial financial returns has attracted great interest for biotechnology in public markets, creating a demand for analysts, venture capitalists, and investment bankers with an understanding of biotechnology-related financial issues.

There is a strong need for specialists who can help with development and manufacturing processes. Engineers with skills in water purification, brewery design and operation, product packaging, and electrical and software design can all find roles in biotechnology production. Bioinformatics research likewise requires individuals with proven computer programming abilities who can apply their skills to biotechnology problems. There is also a strong need for branding and communications professionals. The significant negative impact of patent expirations on sales creates a great opportunity for individuals who can successfully develop strong brand positions for pioneers to help them sustain sales following patent expiration. The volatile nature of R&D and funding also necessitates carefully crafted communications with a diverse audience.

While a majority of individuals involved in research and development have advanced degrees in the sciences, there are numerous opportunities for individuals without scientific degrees. For example, the career path of Kevin Sharer took him from being chief engineer on a nuclear submarine to working at GE and MCI before becoming president and CEO of Amgen, a leader in the biotechnology industry. Prior to joining Amgen, his closest prior exposure to biotechnology was high school biology and college chemistry.

Regardless of educational background or expertise, the key to a career in biotechnology is to appreciate the unique challenges faced by biotechnology companies. By understanding the factors influencing biotechnology companies it is possible to select opportunities and position oneself for a rewarding career.

PH.D., MBA, OR BOTH?

A common question asked by students and business professionals seeking to enter the biotechnology industry is whether a Ph.D. or MBA would improve their career prospects. There are numerous opportunities for individuals with either, or neither, degree. The choice of what to study depends largely upon an individual's career interests and the desire and ability to complete advanced studies. The financial risks and rewards also vary by degree attainment and career path.

Research managers should generally have a Ph.D. in science or a medical degree. Other roles such as marketing, accounting, sales, and human resources do not require an advanced degree in science, but benefit from a business background with an understanding of pertinent biotechnology industry issues.

Because biotechnology companies focus on researching and developing products based on advanced scientific principles, Ph.D. degrees are common among researchers and managers alike. Pharmaceutical companies, with a relatively greater involvement in marketing and licensing, require managers with business expertise in addition to those with scientific backgrounds.

One way to assess the career potential of specific degrees is to examine job postings and the credentials of individuals in positions of interest. Many companies list the academic qualifications of their senior management. Job postings also describe necessary qualifications for specific positions. It is also instructive to look at the non-academic requirements in job postings and the career histories of executives.

Individuals interested in obtaining both a Ph.D. and MBA may find that the preferred route is to obtain a Ph.D. first. While both these advanced degrees require dedication and passion to successfully attain, completing a Ph.D. requires sustained dedication and working long hours for more years than MBA programs require. Beyond time spent in lab, students should be passionate and spend most of their waking hours thinking about their research. The need for dedicated focus and intense time demands are reasons why most first year doctoral students are relatively young. It is accordingly far less common to pursue a Ph.D. after an MBA than to follow Ph.D. studies with an MBA (another possible reason is opportunity-cost: salaries for fresh MBA graduates are generally higher than for fresh Ph.D. graduates).

It is also worth considering other educational options. An alternative to a Ph.D. is to pursue a bachelor's or master's degree. A certificate program may likewise be a reasonable alternative to an MBA. An increasing number of schools are also offering blended programs, combining science and management classes. In the final analysis, the range of educational options depends on one's career motivations. A bachelor's or master's degree may be more appropriate for those seeking to expose themselves to science, but who are not willing to spend a half-decade or more intensely focused scientific research.

Enrolling in a Ph.D., master's, or MBA program with the intention of getting a degree but without being passionate about the topic of study is of relatively little value. The preferred route is to follow your passions, hone your craft, and find ways to apply your expertise to problems you find interesting. Skills learned in school will influence early job roles and career opportunities. Later opportunities will be based upon early performance, and skills and experience learned in earlier positions—being passionate about your job greatly facilitates this growth.

VI

Appendices

Appendix A

Internet Resources

NEWS AND INFORMATION

BiotechBlog
http://www.BiotechBlog.com
This blog, managed by *The Business of Biotechnology* author Yali Friedman, covers new commercial, legal, political, and scientific trends in biotechnology.

Biotechnology@Nature.com
http://www.nature.com/biotech/
The Nature Publishing Group publishes several journals that cover developments and issues in biotechnology. This portal page provides a quick overview of all relevant Nature Publishing Group resources in the field of biotechnology.

Drug Wonks
http://www.drugwonks.com
Drug Wonks is the forum for the Center for Medicine in the Public Interest, covering policy affecting biotechnology. This blog also provides valuable perspectives on many important topics by tracking and responding to Op-Eds and other news items

In the Pipeline
http://pipeline.corante.com/
This blog is written by an active pharmaceutical researcher. His unique insights on the back-stories behind industry developments are an excellent resource to develop a better understanding of the business of biotechnology.

Law.com
http://www.law.com
This portal for legal professionals includes many articles on biotechnology issues. Type "biotech" or "biotechnology" in the search box for quick access to articles on biotechnology.

Mars Blog
http://blog.marsdd.com/
This blog is hosted by an incubator in downtown Toronto and covers topics such as emerging science and technology, entrepreneurship and business, and innovation policy. This blog is exemplary in its depth of coverage and the demonstrated desire to ask, and address, fundamental questions.

Patent Baristas
http://www.patentbaristas.com/
This blog is a must-visit for interpretation of new regulations, patent rulings, or other leading case developments.

Pharma Marketing Blog
http://pharmamkting.blogspot.com/
This might better be described as a source of how *not* to market. The content frequently addresses marketing gaffes and controversies. An excellent source of guidance for breaking stories and case studies.

Sciencecareers.org Career Development
http://sciencecareers.sciencemag.org/career_development
A career development magazine that helps early-career scientists explore their career options. Research and non-research careers in academia, industry, and elsewhere are explored and profiled.

United States Regulatory Agencies Unified Biotechnology Website
http://usbiotechreg.nbii.gov/index.asp
This website focuses on the agricultural products of modern biotechnology. A searchable database covers genetically engineered crop plants intended for food or feed that have completed all recommended or required reviews for food, feed, or planting use in the United States.

USDA Agricultural Biotechnology Briefing Room
http://www.ers.usda.gov/Briefing/Biotechnology
This Economic Research Service production provides background and coverage of issues on adoption and economics of biotechnology in farming. Topics include marketing, labeling, and segregation issues associated with genetically modified foods and agricultural biotechnology research and development of biotechnology.

U.S. Department of State: Biotechnology
http://usinfo.state.gov/ei/economic_issues/biotechnology.html
Produced by the Office of International Information Programs, this site presents speeches, articles, and links on biotechnology policy, regulations, and science. An excellent resource for international and political issues in biotechnology.

INVESTING AND COMPETITIVE INTELLIGENCE

Biospace
http://www.biospace.com
This resource for general and company-specific biotechnology industry news excels in its investing section. Prominent features include financial figures, company information, regional profiles, a career center, and breaking news.

Drug Patent Watch
http://www.DrugPatentWatch.com
Information on pharmaceutical drug patent expirations, sales statistics, generic equivalents, patent claims, pharmaceutical sponsors, and more. Yali Friedman is founder of Drug Patent Watch.

Recombinant Capital
http://www.recap.com
Advice and analysis related to the environment for corporate and product development and alliance formation. In addition to several value-added databases, Recombinant Capital also produces Signals Magazine (www.signalsmag.com), featuring biotechnology industry analysis.

INDUSTRY ORGANIZATIONS

Biotechnology Industry Organization
http://www.bio.org

Pharmaceutical Researchers and Manufacturers of America
http://www.phrma.org

Appendix B
Annotated Bibliography

Building Biotechnology: Business, Regulations, Patents, Law, Politics, Science
Yali Friedman
Logos Press, 2008. ISBN: 978-09734676-6-6
An expanded version of *The Business of Biotechnology,* this is the definitive primer on the business of biotechnology. In addition to extra chapters on politics, international business development, and management, *Building Biotechnology* adds numerous real-world examples on topics such as the hidden pitfalls of common operational decisions; practical considerations in selecting business models and funding options; strategies to overcome developmental failures; and, methods to leverage options to strengthen development plans.

SCIENCE

The Billion Dollar Molecule
Barry Werth
Touchstone Books, 1995. ISBN: 0671510576
This book presents a first-hand account of the development of Vertex Pharmaceuticals. Read about the challenges of selecting projects for drug development and the complex interaction of science and business in biotechnology business development.

Biotechnology Journal
Wiley Interscience
http://www.wiley-vch.de/publish/en/journals/alphabeticIndex/2446/
This peer-reviewed journal publishes papers covering novel aspects and methods in all areas of biotechnology, especially those focusing on healthcare, nutrition and technology. Special attention is also paid to the public, legal, ethical and cultural aspects of biotechnological research.

Chemical and Engineering News
American Chemical Society
http://pubs.acs.org/cen/
A weekly magazine, *Chemical and Engineering News* covers many scientific topics in biotechnology. Print subscriptions are free with Society membership and online access is available for non-members.

Genes IX
Benjamin Lewin
Prentice Hall, 2007. ISBN: 0763740632
The textbook of modern molecular biology, *Genes IX* presents current knowledge on the mechanisms of biological processes. The level of discussion is quite advanced; unfamiliar readers may want to complement this with an undergraduate textbook.

Genetic Engineering News
Mary Ann Liebert, Inc.
http://www.genengnews.com
Published 21 times a year, this tabloid-format publication covers the entire bioproduct life cycle, from early-stage R&D, to applied research and bioprocess, through to commercialization, including marketing and regulations. The application scope includes biopharmaceuticals, bio-agriculture, chemicals and enzymes, environmental markets, and emerging biosciences including biodefense, bioenergy, and nanobiotechnology.

Invisible Frontiers
Stephen Hall, James Watson
Oxford University Press, 2002. ISBN: 0195151593
This book tells the story of the race to clone the first human gene, an achievement that led to the formation of Genentech and the birth of biotechnology. Excellent reading for anyone interested in the history and early development of the biotechnology industry.

Modern Drug Discovery
American Chemical Society
http://pubs.acs.org/journals/mdd/
Modern Drug Discovery focuses on emerging trends in drug discovery. A print subscription is free for individuals employed within the drug discovery and/or life science research fields who live in North America, the United Kingdom, or Western Europe.

Nature Biotechnology
Nature Publishing Group
http://www.nature.com/nbt
Nature Biotechnology, a sister publication of the preeminent scientific journal *Nature*, publishes significant application in the pharmaceutical, medical, agricultural, and environmental sciences. Complementing this function, the journal also features analysis of, and commentary on, published research and business, regulatory, and societal activities that influence this research. A regular supplement series, Bioentrepreneur, provides practical advice on the challenges in building a biotechnology company.

INTELLECTUAL PROPERTY AND REGULATION

From Test tube to Patient
Food and Drug Administration. Fourth Edition, January 2006
http://www.fda.gov/fdac/special/testtubetopatient/
One of the FDA's most popular publications, this report tells the story of new drug development in the United States and highlights the consumer protection role of the Center for Drug Evaluation and Research. Articles describe individual procedures in drug development, from laboratory drug testing to clinical trials and post-marketing surveillance.

Guide to U.S. Regulation of Agricultural Biotechnology Products
Pew Initiative on Food and Biotechnology, 2001
http://pewagbiotech.org/resources/issuebriefs/1-regguide.pdf
This report, focusing on agricultural biotechnology, provides a general overview of the U.S. regulations and laws under which biotechnology products are reviewed for health, safety, and environmental impacts.

IP Management in Health and Agricultural Innovation: A Handbook of Best Practices
A. Krattiger, R.T. Mahoney, L. Nelsen, *et al.*
MIHR: Oxford, UK, and PIPRAL Davis, USA
http://www.iphandbook.org/
This rich guide features 153 chapters by more than 200 authors on practical issues in IP management. Website guides are distill key points in unique contexts designed for policymakers, senior administrators, technology transfer managers, and scientists. A companion blog also provides current commentary on IP management issues.

BUSINESS

The Art of the Start
Guy Kawasaki
Portfolio Hardcover, 2004. ISBN: 1591840562
Written by prolific venture capitalist Guy Kawasaki, *The Art of the Start* provides an overview of the important steps in starting a new venture, including important guidance in refining business plans and pitching investors.

First Fruit
Belinda Martineau
McGraw-Hill, 2001. ISBN: 0071360565
This book profiles the development of the Flavr Savr tomato, the first genetically engineered whole food ever brought to market. Read about the scientific challenges of producing value-added genetically modified plants and the process of satisfying regulatory concerns. Interestingly, Flavr Savr tomatoes did not fail in the market due to public resistance, but rather due to management's inexperience in the premium tomato business which left them unable to sell the tomatoes at a profit.

The Golden Helix
Arthur Kornberg
University Science Books, 1996. ISBN: 0935702326
Nobel laureate Arthur Kornberg was originally skeptical of commercial biotechnology, insisting that academic labs were better equipped to research fundamental issues in biology. In this book, Kornberg details his involvement in the development of Alza, a drug delivery firm, and the growth of the biotechnology industry during this time. Countering his initial sentiments, Kornberg concludes that industry, not academia, is where the most productive science takes place.

Journal of Commercial Biotechnology
Palgrave Macmillan
http://www.palgrave-journals.com/jcb
The *Journal of Commercial Biotechnology* aims to deliver a practical understanding of the strategic development and management associated with the commercialization of biotechnology through dissemination and evaluation of the current techniques, strategic thinking, and best practice in all aspects of the subject. Yali Friedman is managing editor of the *Journal of Commercial Biotechnology*.

The Journal of Life Sciences
Burrill & Company and the California Healthcare Institute
http://www.tjols.com
The Journal of Life Sciences is a bi-monthly magazine focusing on "where science and society meet," and presents analysis and commentary about the impact of biotechnology and other biosciences on business, policy, and culture.

Science Lessons: What the Business of Biotech Taught Me About Management
Gordon Binder, Philip Bashe
Harvard Business School Press, 2008. ISBN: 9781591398615
This book offers a rare glimpse into the early development of Amgen, one of the biotechnology industry's leading companies. Gordon Binder served in the increasingly central roles of CFO, CEO, and Chairman of Amgen from 1982 to 2000—a span which covers the launch and growth of Epogen. In addition to tracing the development of Amgen and the biotechnology industry, Binder and Bashe also offer practical management advice and recommendations on tackling common business challenges.

Term Sheets & Valuations
Alex Wilmerding
Aspatore Books, 2006. ISBN: 1587620685
An in-depth look at term sheets. In addition to a section-by-section view of a term sheet, valuations, and guidance, this book includes a sample term sheet with a description of each clause and a discussion of key negotiation points and red flags.

U.S. and Canadian Biotechnology VC Directory
BioAbility, LLC
http://www.bioworld.com/
A comprehensive listing of VC firms investing in biotechnology.

Valuation in the Life Sciences: A Practical Guide
Boris Bogdan and Ralph Villiger
Springer, 2007. ISBN: 9783540455653
This comprehensive source for biotechnology valuation offers a mix of theory and practical examples. A diverse set of valuation examples is covered, helping ground the theory and enabling readers to understand when, and how, to apply valuation methods.

The Use of Biotechnology in U.S. Industries
U.S. Department of Commerce, 2003
http://www.technology.gov/reports/Biotechnology/CD120a_0310.pdf
The first ever in-depth federal government assessment of the development and adoption of biotechnology in industry. This assessment was directed at increasing national policy makers' understanding of the current development and use of biotechnology in U.S. industries, and to assist federal statistical agencies in developing measures and statistics of biotechnology related economic activity.

Glossary

SCIENCE

Absorption, Distribution, Metabolism, Excretion and Toxicology (ADMET): An element of pre-clinical and clinical trials used to measure the effects of a drug on animal and human physiology.

Amino acid: Building block of proteins. Proteins consist of amino acids linked end-to-end. There are 20 different amino acid molecules that make up proteins. The DNA sequence that codes for a gene dictates the order of amino acids in a given protein.

Antibiotic: A chemical substance that can kill or inhibit the growth of a microorganism.

Antibody: Immune system protein produced by humans and higher animals to recognize and neutralize bacteria, viruses, cancerous cells, and other foreign compounds.

Antisense: A natural or synthetic DNA or RNA molecule that specifically binds with messenger RNA to selectively inhibit expression of a single gene.

Applied research: Aimed at gaining knowledge or understanding to determine the means by which a specific recognized need may be met. Applied research builds upon the discoveries of basic research to enable commercialization.

***Bacillus thuringiensis*:** A naturally occurring bacteria that produces Bt toxin, a protein that is toxic to certain kinds of insects. The Bt toxin gene has been genetically engineered into corn and cotton plants to reduce the need for chemical pesticides.

Bacteriophage: Naturally-occuring type of virus that only infect bacteria.

Base: A key component of DNA and RNA molecules. Four different bases are found in DNA: adenine (A), cytosine (C), guanine (G) and thymine (T). In RNA, uracil (U) substitutes for thymine.

Basic research: Aimed at gaining more comprehensive knowledge or understanding of the subject under study, without specific applications in mind. Basic research advances scientific knowledge but does not have specific immediate commercial objectives, although it may be in fields of present or potential commercial interest.

Biofuel: Fuels such as ethanol and diesel produced from sugars, vegetable oils, or other organic matter using biotechnology methods.

Bioinformatics: The application of information technology to manage and analyze the vast amounts of data generated from biological research.

Bioleaching: The use of plants to extract heavy metals from soils.

Bioremediation: The use of biological systems, usually microorganisms, to decompose or sequester toxic and unwanted substances in the environment.

Biotechnology: The application of molecular biology for useful purposes.

Blue biotechnology: A seldom-used term referring to marine and aquatic applications of biotechnology.

Chromosome: The DNA-protein complexes that contain all the genes in a cell.

Cloning: The process of making an identical copy of something. Often used in reference to copying animals, it may also refer to creating copies of DNA fragments, individual cells, or plants.

Codon: A sequence of three DNA or RNA bases that specifies an amino acid in the synthesis of a protein.

Combinatorial chemistry: A product discovery technique that uses robotics and parallel chemical reactions to generate and screen as many as several million molecules with similar structures in order to find chemical molecules with desired properties.

Cytochrome p450: A set of enzymes involved in chemical modification and degradation of chemicals including drugs and other foreign compounds.

Data mining: Using computers to analyze masses of information to discover trends and patterns.

Diagnostic: A product used for the diagnosis of a disease or medical condition.

DNA (deoxyribonucleic acid): The primary source of genetic information in cells. DNA is comprised of nucleotides and is composed of two strands wound around each other, called the double helix.

DNA fingerprinting: A DNA analysis method that measures genetic variation among individuals. This technology is often used as a forensic tool to detect differences or similarities in blood and tissue samples at crime scenes.

DNA sequencing: The process of determining the exact order of bases in a segment of DNA.

Double-blind: An experimental protocol whereby neither the experimental subjects nor the administrators know whether a drug or placebo is being administered. Double-blind protocols are used to eliminate bias.

Drug delivery: The process by which a formulated drug is administered to the patient.

Drug development: The process of taking a lead compound, demonstrating it to be safe and effective for use in humans, and preparing it for commercial-scale manufacture.

Enzyme: A functional protein that catalyzes (speeds up) a chemical reaction. Enzymes control the rate of naturally occurring metabolic processes such as those necessary for growth and reproduction.

Escherichia coli (*E. coli*): A common gut bacteria that is a workhorse and model organism for molecular biology.

Excipient: An inactive ingredient (there are no absolutely inert excipients) added to a drug to give it a pill form or otherwise aid in delivery.

Expression: A highly specific process in which a gene is switched on at a certain time and its encoded protein is synthesized, resulting in the manifestation of a characteristic that is specified by a gene. Genetic predispositions to disease arise when a person carries the gene for a disease but it is not expressed.

False negative: An experimental outcome that incorrectly yields a negative result. False negatives can complicate disease diagnosis.

False positive: An experimental outcome that incorrectly yields a positive result. False positives can frustrate assessing the performance of lead compounds.

Fermentation: Technically the process of breaking complex organic substances into simpler ones, such as conversion of sugars into alcohols, acetone, or lactic acid. Also refers to any large-scale cultivation of microbes or other single cells (e.g., for drug production).

Functional genomics: The use of biological experiments and genetic correlations to establish what each gene does, how it is regulated, and how it interacts with other genes.

Functional foods: Foods containing compounds with beneficial health effects beyond those provided by the basic nutrients, minerals, and vitamins.

Gene: The fundamental unit of heredity, a segment of DNA which encodes a defined biochemical function. Some genes direct the synthesis of proteins, while others have regulatory functions.

Gene expression: The production of a gene product—generally defined as the synthesis of an encoded protein.

Gene splicing: Splicing a gene from one segment of DNA into another. Commonly used to insert foreign genes into bacteria for analysis, or to insert foreign genes into bacteria or other organisms for genetic modification or to produce and harvest large quantities of specific proteins.

Gene therapy: The replacement of a defective gene in a person or organism suffering from a genetic disease.

Genetic code: The language in which DNA's instructions are written. The genetic code consists of triplets of nucleotides (codons), with each triplet corresponding to one amino acid in a protein structure, or a signal to start or stop protein production.

Genetic disorder: A condition or mutation that results from one or more defective genes.

Genetic engineering: The manipulation of genes to create heritable changes in biological organisms and products that are useful to people, living things, or the environment.

Genetic predisposition: A susceptibility to disease that is related to a genetic condition, which may or may not result in actual development of the disease.

Genetic screening: The use of a specific biological test to screen for inherited diseases or medical conditions.

Genome: The sum of an organism's genes.

Genomics: The study of genes and their function.

Good manufacturing practice (GMP): Guidelines ensuring the quality and purity of chemical products that are intended for use in pharmaceutical applications, and controls ensuring that methods and facilities used for production, processing, packaging, and storage result in drugs with consistent and sufficient quality, purity, and activity.

Gray biotechnology: A seldom used term for industrial applications of biotechnology. More commonly referred to as white biotechnology.

Green biotechnology: The use of biotechnology for agricultural applications.

Human Genome Project: The international research effort which identified and located the full sequence of bases in the human genome.

Incidence: measure of the rate of new occurrences of a disease or condition in a population.

Immune system: The cells, biological substances (such as antibodies), and cellular activities that work together to recognize foreign substances and provide resistance to disease.

***In silico* (in computer):** Computer-based predictions that can complement *in vitro* and *in vivo* procedures.

***In vitro* (in glass):** Experimental procedures carried out in testtubes, beakers, etc.

***In vivo* (in the living body):** Experimental procedures carried out on living cell lines or in living animals.

Lead compound: In pre-clinical development and clinical trials, a potential drug being tested for safety and efficacy.

Liposome: An artificial membrane. Can be used to encapsulate drugs and aid in drug delivery.

Microarray: A tool that permits the identification of DNA samples and examination of gene expression in individual tissues and different conditions.

Monoclonal antibody: A synthetic immune system protein that recognizes a single target. Polyclonal antibodies recognize several related targets.

Molecular evolution: The process of making discrete changes in genes to improve the functional characteristics of proteins and enzymes.

Molecular farming: Using biotechnology to produce useful products from domesticated plants and animals.

mRNA (messenger RNA): A ribonucleic acid molecule that transmits genetic information from DNA to the protein synthesis machinery in cells, where it directs protein synthesis.

Mutant: A cell or organism harboring one or more mutated genes.

Mutation: A change in the base sequence of a gene that results in it not performing its normal task.

Nanotechnology: A technology field focusing on materials at sizes measured in billionths of a meter.

Nucleotide: One of the structural components, or building blocks, of DNA and RNA. A nucleotide consists of a base plus one molecule of sugar and phosphoric acid.

Oncogenic: Viruses, chemicals, genes, proteins, etc. that cause the formation of tumors.

Pathogen: A disease-causing organism.

Personalized medicine: The practice of medicine in which therapies are developed for and directed at the patients most likely to benefit from them.

Pharmacogenetics: Examination of the differences in drug response between individuals—one drug, many genomes.

Pharmacogenomics: Examination of differences in how one person responds to different drugs—many drugs, one genome.

Pharming: The process of farming genetically engineered animals and plants to produce drugs.

Placebo: A mock-treatment used in single-blind or double-blind experiments to eliminate bias from experiment subjects or administrators, respectively.

Platform technology: A technique or tool that enables a range of scientific investigations. Examples include combinatorial chemistry for producing novel compounds, microarrays for gene expression analysis, and bioinformatics programs for data assembly and analysis.

Polymerase Chain Reaction (PCR): A method to produce sufficient DNA for analysis from a very small amount of DNA.

Prevalence: measure of how commonly a disease or condition occurs in a population.

Prion: A naturally occurring protein that can be converted into a disease-causing form. Prion diseases can be transmitted in the absence of DNA or RNA.

Promoter: A DNA sequence preceding a gene that contains regulatory sequences influencing the expression of the gene.

Proof-of-principle: Demonstration of the commercial potential of a discovery or invention.

Protein: A long-chain molecule comprised of amino acids that folds into a complex three-dimensional structure. The type and order of the amino acids in a protein is specified by the nucleotide sequence of the gene that codes for the protein. The structure of a protein determines its function.

Proteomics: The study of the protein profile of each cell type, protein differences between healthy and diseased states, and the function of, and interaction among, proteins.

Rational drug design: Using the known three-dimensional structure of a molecule, usually a protein, to design a drug that will bind have a therapeutic effect on it.

Recombinant DNA: The DNA formed by combining segments of DNA from different sources.

Red biotechnology: The use of biotechnology for therapeutic applications.

Reformulation: Altering an established drug's formulation or delivery method to yield improvements in safety or efficacy.

Repurposing: Finding new indications for approved drugs.

Restriction enzyme: A protein that cuts DNA molecules at specific sites, dictated by the nucleotide sequence.

Retrovirus: A type of virus that reproduces by converting RNA into DNA.

Single Nucleotide Polymorphism (SNP): A single base difference in the sequence of a gene which alters the structure and function of the gene product.

RNA (ribonucleic acid): A nucleic acid, similar to DNA, which has roles in gene expression.

RNA interference: Using antisense techniques to selectively inhibit expression of a gene.

Stem cell: An undifferentiated cell that can multiply and become any sort of cell in the body.

Telomere: The tip of a chromosome. Telomeres are involved in the replication and stability of chromosomes.

Tissue engineering: The production of natural or synthetic organs and tissues that can be implanted as fully functional units or may develop to perform necessary functions following implantation.

Transcription: The synthesis of an mRNA molecule as a copy of a gene. In gene expression, transcription precedes translation.

Translation: The synthesis of a protein based on the nucleotide sequence of an mRNA molecule, which corresponds to the sequence of a gene.

Transgenic: An organism with one or more genes that have been transferred to it from another organism.

Vaccine: A preparation of either whole disease-causing organisms (killed or weakened) or parts of such organisms, used to confer immunity against the disease that the organisms cause. Vaccine preparations can be natural, synthetic, or derived by recombinant DNA technology.

White biotechnology: The use of biotechnology for industrial applications.

X-ray crystallography: An essential technique for determining the three-dimensional structure of biological molecules.

Xenotransplantation: Transplanting a foreign tissue into another species.

LEGAL

Claim: A comprehensive and precise description that defines the scope of an invention.

Compulsory license: A license in which a government forces the holder of a patent or other exclusive right to grant use to the state or others. Authorized under World Trade Organization provisions to enable countries to produce generic versions of patented drugs in the event of a health crisis.

Continuation: A filing, while a patent is active, which contains additions or changes to the previous claims.

Copyright: The exclusive legal right to publish, perform, display, or distribute an original work.

Divisional patent: A patent that covers the same specification as a previous (parent) patent, but claims a different invention.

Ex parte: A legal proceeding where only one party is represented. Patent prosecution is an *ex parte* procedure.

Experimental use: The practice of a patented invention solely with intention of experimentation or perfection of the invention.

Evergreening: The practice of launching new formulations, combinations, delivery methods, and indications for drugs facing patent expiration to effectively increase the duration of patent-protected sales.

Filing date: The date on which a complete patent application is received by the Patent and Trademark Office.

Freedom to operate: The absence of intellectual property and regulatory impediments (which may require patent license and passage of enabling laws) to commercialization.

***Inter partes* reexamination:** A method by which third parties challenge the validity of a patent on the grounds of prior art publication without resorting to litigation.

Interference: When two or more patent applications or issued patents claim the same invention.

License: An agreement whereby one party gains access to another's technology (e.g. a patent license).

Non-disclosure agreement (NDA): An agreement, common between companies and their contractors and partners, which allows a company to share protected information while preventing its release.

Office action: A formal response by a patent examiner regarding a patent application or amendment.

Patent: A description of an invention. Patents contain one or more claims that describe the subject matter covered in sufficient detail to permit skilled experts to practice an invention, and grant the right to exclude others from practicing an invention.

Patent agent: An individual with technical training who is capable of representing an inventor in patent prosecution.

Patent attorney: An individual with legal training in patent law who is capable of representing an inventor in patent prosecution and litigation.

Patent pool: An agreement between two or more patent owners to license one or more of their patents to one another or third parties.

Patent term adjustment: Provisions to adjust patent term to provide restoration for U.S. patent and trademark office delays.

Patent term extension: Provisions to extend patent term to account for time spent waiting for FDA approval.

Prior art: Public knowledge that exists in a field; all previously issued patents, publications, public announcements, or knowledge that bear on the invention claimed in a patent application.

Prosecution: The process by which an inventor engages with the patent office to obtain a patent and determine the scope of its claims.

Provisional patent application: A preliminary patent application filed without a formal patent claim, oath or declaration, or any prior art statement. It provides the means to establish an early effective filing date in a subsequent non-provisional patent application and allows the term "Patent Pending" to be applied.

Reach-through claim: A patent claim to rights to royalties from, or rights to use, drugs or other physical or intellectual property produced using the patent

Submarine patent: A patent that emerges after it has unknowingly been infringed upon.

Trade secret: Knowledge and information that is not generally known to the industry. Examples include customer lists, business plans, and manufacturing methods.

Trademark: A registered name, word, symbol, or device identifying a company's products or services.

REGULATORY

Abbreviated New Drug Application (ANDA): A simplified submission permitted for a generic version of an approved drug.

Accelerated approval: A process to make products for life threatening diseases available on the market prior to formal demonstration of benefit. Uses surrogate markers—indirect measures of efficacy—and requires continued testing to confirm efficacy.

Action letter: An official FDA communication that informs the sponsor of an NDA or BLA of a decision by the agency. An approval letter allows commercial marketing of the product.

Authorized generic: Drugs produced by branded companies and marketed under a private label to compete with other generic drugs.

Bayh-Dole Act: Provides the statutory basis and framework for federal technology transfer activities, including patenting and licensing federally funded inventions to commercial ventures.

Bioequivalence: Demonstration that a generic drug has the same chemical and biological properties as its pioneer counterpart.

Biologic: Medicine made by biological processes rather than by chemical synthesis or extraction. Biologics typify biotechnology-derived drugs. Contrast with small-molecule drugs.

Biologics License Application (BLA): Application filed with the FDA Center for Biologics Evaluation and Research (CDER) for approval to market a biologic drug.

Biosimilar: A generic biologic drug that is "similar but not identical" to a pioneer drug.

Brand-name drug: The original, often patented, version of a drug. Contrast with generic drugs.

Clinical pharmacology study: Clinical trial designed to determine the absorption, distribution, metabolism, elimination, and toxicity (ADMET) of a drug.

Clinical trial: A human study designed to measure the safety and efficacy of a new drug.

Current good manufacturing practices (cGMP): Regulatory practices to ensure safety and consistency of manufacturing processes.

Exclusivity: A temporary FDA-granted monopoly, distinct from patent or other intellectual property protection. Exclusivity may be granted for developing drugs for rare diseases, novel drugs, conducting pediatric clinical trials, or successfully challenging invalid patents.

Fast track: A process for interacting with the FDA during drug development, intended for drugs to treat serious or life threatening conditions that demonstrate the potential to address an unmet medical need.

First-in-man study: Phase I trial primarily concerned with establishing the safety of a compound.

Follow-on biologic: An FDA term for a biologic drug that is similar to an existing biologic.

Generic drug: The version of an approved drug produced by a competitor after a pioneer firm's patents expire.

Hatch-Waxman safe harbor: A research-use exemption stemming from the Hatch-Waxman Act which exempts from infringement the use of patented inventions in preparation for submitting drug applications.

Hatch-Waxman Act: Contains provisions to foster the development of generic drugs and support pioneer drug development.

Indication: A use for which a specific drug is approved by the FDA.

Institutional Review Board (IRB): An independent committee of scientists, physicians, and lay people that oversees clinical trials.

Investigational New Drug (IND): An application to pursue clinical trials with an experimental drug that has passed pre-clinical development.

March-in rights: A stipulation of the Bayh-Dole Act enabling the government to request and potentially require issuance of a license to a patent, which was developed with federal funding, to another party.

Named Patient Program: European compassionate use program, enabling limited distribution of drugs prior to approval.

New Drug Application (NDA): Application filed with the FDA Center for Drug Evaluation and Research (CDER) for approval to market a small-molecule drug.

Off-label use: Use of a drug not in accordance with FDA-approved uses or drug labeling. Physicians are free to prescribe drugs for off-label uses.

Orange Book: Also known as *Approved Drug Products with Therapeutic Equivalence Evaluations*, the *Orange Book* contains detailed information on all approved drugs and must list all extant patents.

Orphan Drug: A drug that treats a disease affecting fewer than 200,000 Americans or for which there is no reasonable expectation that the cost of research and development will be recovered from sales in the United States. The Orphan Drug Act provides special incentives for producers of orphan drugs.

OTC-switch: The process of gaining approval to sell a drug over the counter, which may grant 3 years exclusivity for the over-the-counter market.

Over the counter (OTC): Selling a drug without a prescription. Requires evidence that patients can self-diagnose and use the drug safely without physician supervision.

Phase I: Clinical trial designed primarily to determine the safety of an experimental drug.

Phase II: Clinical trial that evaluates an experimental drug's safety, assesses side effects, and establishes dosage guidelines.

Phase III: Clinical trial designed to assess the safety and effectiveness of an experimental drug. Success in Phase III trials can lead to marketing approval.

Phase IV: Post-approval clinical trials used to monitor safety and efficacy or examine additional applications of drugs.

Pioneer (brand-name) drug: The patented version of a drug. Contrast with generic drugs, the competing versions produced when pioneer patents expire.

Pre-clinical studies: Studies that test a drug on animals and non-human test systems. Safety information from such studies is used to support an investigational new drug application (IND).

Reverse payment: A payment from a branded drug company to a generic drug company to delay launch of a generic drug.

Salami slicing: Filing for multiple orphan drug designations on the same drug.

Small-molecule drug: A drug produced using defined chemical synthesis or extraction. Contrast with biologics, drugs produced by biological processes.

Surrogate marker: An indirect measure of effectiveness, such as a laboratory test or tumor shrinkage, used to show a strong potential for effectiveness in accelerated drug approval.

COMMERCIAL

Accredited investor: A type of investor, largely defined by their wealth, permitted to invest in high-risk investments.

Acquisition: Appropriation of the controlling interests of one company by another.

Alliance: Agreement between two or more companies to cooperate in some way.

Angel investor: Wealthy individual who personally provides startup capital to very young companies to help them grow.

Barrier to entry: A condition that makes it difficult for competitors to enter the market (e.g., patent, trademark, high up-front capital requirements).

Blockbuster: Drug with $1 billion or more in sales.

Board of directors: A group legally charged with the responsibility to protect the interests of a company and its shareholders.

Bootstrap: Starting a business with little or no external funding.

Bridge loan: A short-term, high-interest, loan provided to companies in dire need of cash.

Burn rate: The rate at which an unprofitable company is going through its available cash reserves.

Business model: A description of a company's purpose, commercial offerings, strategies, organizational structure, operational processes, etc. Often confused with business plan, below.

Business plan: A formal statement of a company's goals and the plan for reaching those goals.

Comparable: A valuation technique based on analogy to similar companies or products.

Competitive advantage: An advantage that a firm has relative to competing firms; may be in the form of intellectual property, expertise, partnerships, assets, etc.

Controlling interest: Ownership of more than 50 percent of a company's voting shares.

Convertible: Securities (usually bonds or preferred shares) that can be converted into common stock.

Cooperative research and development agreement (CRADA): An agreement enabling federally funded laboratories to perform for-profit contract work for commercial firms.

Corporate inversion: Formation of a parent corporation of a U.S. company in a country with little or no corporation tax, and structuring a U.S. subsidiary to manage U.S. sales. This scheme enables foreign sales to be taxed in local markets only, and not be taxed in the U.S.

Cross-licensing: An agreement in which two or more firms with competing and similar technologies strike a deal to reduce the need for legal actions to clarify who is to profit from applications of the technology.

Dilution: The decrease in relative ownership among existing investors as additional shares are issued.

Discounted cash flow (DCF): A valuation technique that attempts to consider future events and determine a present value for a product or project based on the variety of outcomes.

Discovery rights: Selling only research findings while retaining rights to knowledge discovered in the course of research and development.

Down round: A financing event in which a company is valued lower than it was previously.

Due diligence: The process by which research is conducted to determine the value of an investment, licensing agreement, merger, or other similar activity.

Earnings strippings: A potential result from a corporate inversion where payments from a U.S. subsidiary to a foreign parent are used as deductions against U.S. taxes.

Elevator pitch: A short summary—typically less than two-minutes—used to quickly describe a business to investors

Dumb money: Funding from investors who cannot provide additional benefits such as guidance or networking.

Equity dilution: The dilution of the equity stakes of founders and early investors by subsequent investments.

Equity investment: An investment purchasing partial ownership of a company.

Exit: The means by which investors gain a return on their investment, commonly through sale of shares in public markets or acquisition by another company.

Free cash flow: The amount of cash available to a company after all expenses have been paid.

Friends and family: A term for investments which often help start a company, and are typically made by unaccredited investors.

Incubator: A facility offering space and shared services and facilities to early-stage companies.

Initial Public Offering (IPO): The initial sale of shares of a private company in public markets, turning it into a publicly-traded company.

Institutional support : Esteem granted to companies by their affiliation with highly regarded partners, financiers, and other affiliates.

Intellectual property: Intangible assets such as patents, trade secrets, trade names, etc.

License: An agreement to grant rights to a patent or tangible subject.

Market segmentation: The division of a market into distinct groups of buyers or decision makers.

Medicaid: Government-subsidized healthcare coverage for individuals with low incomes and limited resources.

Medical tourism: Travelling for medical treatment. Often motivated by cost-savings, local waiting lists and expertise availability, or local regulatory restrictions.

Medicare: Government-subsidized healthcare for individuals 65 years of age and older, some disabled people under 65 years of age, and people with permanent kidney failure.

Merger: The formal combination of two companies into one entity. Often used to refer to acquisitions. A merger can be distinguished from consolidation, in which a new separate entity is created.

Mezzanine funding: Funding that generally leads to liquidity (IPO or merger) or commercial launch and eventual profitability.

Milestone: The completion of a specified phase in product development. Investors and alliance partners may use milestones to establish a timeline for incremental investments or payments.

Offshoring: The relocation of business processes from one country to another.

Options-pricing: A valuation technique that analyzes the value of discrete operational paths.

Outsourcing: The execution and management of selected operations by outside parties.

Parallel trade: The trade of products between countries without permission of the intellectual property owner. Often used to capitalize on inter-country price differences.

Pharmacoeconomics: Study of the cost-benefit ratios of drugs.

PIPE (Private Investment in Public Equity): Purchase of discounted shares in a public company in which payment goes directly to the company rather than to existing shareholders.

Preferred stock: A convertible offering that cannot be sold until it is converted to common stock.

Price elasticity: A measure of the change in demand resulting from a change in price of a product or service. Low price elasticity indicates little change in demand; high elasticity indicates a relatively large change in demand.

Price/earnings (P/E) ratio: A rudimentary technique to determine the value of a company, or relative value of several companies, by comparing the share price to annual earnings.

Private equity: In contrast to owning shares in a public company, private equity is ownership in a private company.

Proxy fight: A process by which shareholders can vote for corporate changes. May be used to appoint new directors or replace senior management.

Royalty: The payment of a percentage of sales as compensation to product developers, patent licensors, or even investors.

Ratchet: An anti-dilution provision where an investor is granted additional shares of stock without charge if the company later sells the shares at a lower price.

Return on Investment (ROI): Profit (or loss) on an investment, often expressed as a percentage.

Reverse merger: The merger of a private company with a "shell" company, rendering the private company public.

SBIR (Small Business Innovation Research): A funding program that encourages small business to explore their technological potential and provides the incentive to profit from its commercialization.

Scientific Advisory Board (SAB): A group of esteemed scientists and business professionals, independent from management, which provides objective feedback and guidance on a company's progress and goals.

Seed financing: Capital furnished to prove the feasibility of a concept or invention.

Secondary offering: A public or private share offering subsequent to an initial public offering.

Series A/B/C/D: Venture funding stages that fund product development and early commercial launch activities.

Shell: A public company with few or no assets that may be the remnant of a bankruptcy or asset sale. Used in reverse mergers to enable a private company to become public.

Smart money: Funding from investors who are able to contribute guidance, networking, or other benefits.

Special purpose acquisition company (SPAC): A public company formed with the intent of engaging in a reverse merger with a private company.

Spin-off: Separating a smaller unit from an established company, permitting each company to retain focus while shielding the parent from risk and granting the spin-off the administrative benefits of small size.

STTR (Small Business Technology Transfer): A funding program that encourages public/private sector partnership in order to develop new technologies and profit from their commercialization.

Targeted marketing: The alignment of marketing efforts with the benefits sought by individual market segments.

Tax arbitrage: Location of specific operations in countries and regions with favorable tax treatment for those operations.

Technology transfer: The transfer of discoveries made by basic research institutions, such as universities and government laboratories, to the commercial sector for development into useful products and services.

Venture capital: Money invested by venture capitalists in startup companies in exchange for equity.

Venture capitalist: An individual who invests in start-up companies with the intent of making a large return on investment.

Virtual company: Firms that outsource all or most of the elements of research, development, and marketing.

Index

About the Author

Yali Friedman is managing editor of the *Journal of Commercial Biotechnology* and serves on the science advisory board of Chakra Biotech and the editorial advisory boards of the *Biotechnology Journal, Journal of Medical Marketing* and *Open Biotechnology Journal*. He regularly guest-lectures for biotechnology education programs, teaching classes on the business of biotechnology, and has written and given talks on diverse topics such as biotechnology entrepreneurship, strategies to cope with a lack of management talent and capital when developing companies outside of established hubs, and new paradigms in technology-based economic development.

Yali also has a long history in biotechnology media, having created a *Forbes* "Best of the Web"-rated web site on the biotechnology industry for a NY Times company and managed it for many years. His other projects include the Student Guide to DNA Based Computers, sponsored by FUJI Television, Biotech-Blog.com, and DrugPatentWatch.com, a pharmaceutical industry competitive intelligence service.

Yali can be contacted at *info@thinkbiotech.com*.

Related titles from Logos Press
www.logos-press.com/books

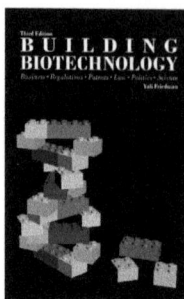

Building Biotechnology
An expanded version of *The Business of Biotechnology,* this is the definitive primer on the business of biotechnology
Softcover: 978-09734676-6-6
Hardcover: 978-09734676-5-9

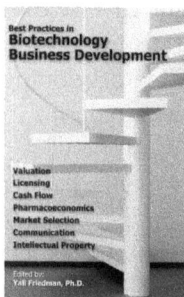

Best Practices in Biotechnology Business Development
Eleven chapters from biotechnology industry experts
ISBN: 978-09734676-0-4

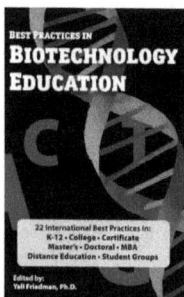

Best Practices in Biotechnology Education
22 chapters on programs from 5 countries
ISBN: 978-09734676-7-3